©2006 by Noel Brodsky

All rights reserved. No part of this publication may be reproduced by any means whatsoever without written permission of the author, with the exception of brief quotations used in critical articles or reviews.

Publisher's Cataloging-in-Publication Data

Brodsky, Noel
A Short Drive Through the 21st Century: The Future of Energy, Trade and Demographics

Bibliography: p.
1. Economic Forecast. 2. Energy Markets. 3. International Trade. 4 International Finance. 5. Demographics. I. Title

Library of Congress Control Number: 2006908017

ISBN 978-1-84728-550-8

Cover design by Bogdan Stanciu

Published by Fringe Lunatic Press, Savoy, IL USA
Printed and distributed by Lulu.com

A Short Drive Through the 21st Century: The Future of Energy, Trade and Demographics

Noel Brodsky, Ph.D.
Department of Economics,
Eastern Illinois University
Charleston, IL

Acknowledgments

This book was improved by many people who gave their energy in helping me write this book. There are some I have remembered to thank here, in no particular order.
Early readers of the drafts of this book were Tamas Forgacs, Sue Mauck and Brian Holding, who made many suggestions that I tried to incorporate into the text. Charles A. Moore offered both early and late editorial help, and my daughter, Ariana provided one of the copy edit passes. My wife, Ann, provided a final editorial review. James R. Bruehler, my colleague at Eastern Illinois University, not only gave me input to economic ideas in the book, but also provided editorial assistance. Joseph Finnerty also helped clean up some of the economic and financial issues presented. Bogdan Stanciu did the graphic design work for the cover.
My wife, Ann, and my daughters, Ariana, Serena and Alyssa, had to put up with me for nearly two years while writing this book. No doubt I regularly bored them with the ideas and analysis I was coming up with, but they seem to take it in stride, though one escaped to Germany on a one year Rotary exchange. She's one the one that got stuck with a late copy edit pass (see above). Peter B. Alexander also had to hear about all of these crazy ideas, often the day I came up with them. While I was writing, he vacationed to Mexico (twice!) and to Alaska, which I'm sure were insufficient breaks from me.
My colleagues in the Department of Economics at Eastern Illinois University also had to deal with me, and they too, seemed to han-

dle it well. The University also provided me with a sabbatical assignment to allow me to do the bulk of the work associated with writing the text.

Any errors or omissions rest with me, as I often failed to do what I was told.

For the Sweet Pea, the Pumpkin and the Red Pepper - it's their future.

CONTENTS

Acknowledgments   iii

1 Preview: Oil, Trade And Demography   1

2 Oil Is Cheap   5
   2.1 A tale of three oil price histories 7
   2.2 How oil got started 13
   2.3 Various forms of energy 20
       2.3.1 Petroleum alternatives you should know . . . 20
       2.3.2 Some non-petroleum alternatives to consider . 22
   2.4 The effect energy has on other markets 28
       2.4.1 High oil prices make oil substitutes viable . . 28
   2.5 High energy prices hurt some industries 33
       2.5.1 The damaging effects of high oil prices 33
       2.5.2 High energy prices have damaging effects too 34
   2.6 Environmental issues to consider 36
   2.7 A quick comment on water issues 39
   2.8 Want to find out more about energy? 41

3 A World Divided: Trade And Finance   43
   3.1 How international trade might change 43
   3.2 Free trade is nice 44

3.3		Resistance to international trade	49
3.4		Globalization and anti-globalization	52
3.5		China, India and Brazil	53
	3.5.1	China should be a successful economy	53
	3.5.2	India has an uncertain future	54
	3.5.3	Brazil has a geographic advantage	56
3.6		US financial issues and a risky US dollar	56
3.7		World financial flows are important too	59
3.8		What to expect in US economic growth	63
3.9		Some resources for international trade	65

4 The Demographic Future — 69
4.1	An aging population	70
4.2	A rising dependency ratio	74
4.3	The damage of HIV/AIDS	78
4.4	China and the missing girls	79
4.5	Aging societies change what's in demand	81
4.6	Carrying capacity: resources at the limit	82
4.7	Demographic change influences markets	86
4.8	Places to look for demographics	87

5 Mix It Up: A Future With All In Play — 91
5.1	A first look - the future as you saw it		93
5.2	A new future: trade in an aging world		99
	5.2.1	Will the 21st century be Africa's?	100
5.3	The environment, energy and technology		102
5.4	A fall in the US dollar in our future		107
5.5	Where to find other views of the future		110

CHAPTER 1

Preview: Oil, Trade And Demography

Over the past decade I've noticed something about my friends and acquaintances: they are worried about the price of oil. At times, they seem downright terrified; the price of oil will rise without bound, or worse, we will run out of oil entirely, and in short order. Perhaps these conversations take place because I'm an economist and folks are looking for any bit of information to help them deal with it. I think it is deeper than that.

Most folks have a strong sense of what is going on around them. They know that oil prices are moving, and they are profoundly affected by sharp increases in the price of oil, especially one of its chief derivatives, gasoline. High gasoline prices mean less to spend on other things; it means a vacation much closer to home or none at all. High heating oil prices mean a colder home, and perhaps fewer presents during the holidays. If prices keep rising, living could become much less pleasant.

Worse, the television news (and newspapers) publish doomsday predictions about the world running out of oil. Whenever oil supply disruptions happen, and the price of oil spikes, some "expert" comes forth and says "expect oil prices to continue to rise for the foreseeable future." Or worse, "we will run out of oil in 30 years, and then what are we going to do?" Heavy stuff. No wonder oil prices are the topic of conversation.

Let's get something straight right away: we are not going to run out of oil anytime soon. It just isn't going to happen. Not in 30 years, not in a century, probably not in two centuries. Will the price of oil rise? I can't be as sure about that, but the analysis presented in this book suggests that on average it probably will remain in inflation-adjusted terms about the same as it has been for the last 140 years. The average price is around $25 a barrel (in 2005 US$). No kidding. That's what you should expect, on average. Oil's price does appear to be in a more volatile period, so repeated price spikes may well be part of the future, but evidence suggests the price will return to its mean.

But the future is so much more than oil prices. Sure, energy prices may be a large part of the driving forces that make up what the future will look like, but there are at least two more issues that need to be addressed: international trade and demographic change. This book's premise is that there are 3 core issues that will affect the United States economy and that of the world in the next 100 years. These core issues are, broadly, highly variable energy costs, population aging, and a decline in the level of world trade. Each of these issues have their own inertia, with underlying economics behind them, and their interactions with each other need to be explored to better understand what the future might look like. By no means do I intend to suggest that these are the only things that matter, or that they are the only things that will happen. Rather, these three issues can be reasonably expected to be part of the century, and with proper analysis they should provide a framework for much of what may happen in this century.

We explore each of the three issues first by establishing the issues themselves, why they are likely to occur, and how the market process is likely to deal with them. Since each has its own investment implications, there is some discussion of what might be good investments to make based on the results of the issue surfacing. There is also a discussion of likely triggers that might help us know when an issue has become important.

Even though this book is about the future, it is not intended to be a prognostication. It is more about taking a few simple forecasts and exploring them using the tools of economics to place them in a larger context. If these simple forecasts are right, it should be straightforward to see how they affect both the markets of interest and related markets. Since most inter-market relationships are known, it should be possible to make proper long-term investment decisions.

In the last chapter of this book, I offer two alternative long range forecasts of the future. The first scenario offers you what happens based on what you believe right now - that oil prices will continue to rise. The second scenario is built off of the analysis presented in this book - oil prices remain, on average, about the same, though with repeated price spikes. The price spikes do some damage, but the real driving forces become international trade and demographic change. The structure of the world changes, but with the right preparation for a world with an older population, the change is manageable. You could skip the middle chapters and read the last chapter first, but you may find it difficult to follow some of the ideas without the supporting details provided in the earlier chapters.

Economic history has taught us that prices rise and fall over time. Markets are in a constant state of flux, as they seek new prices and quantities. No matter what direction they head in, if you are in the right position you can profit from the movement. The trick is figuring out what position to be in, and when to change your market position. This book should help you with your market decisions.

While I use economic analysis to explore these issues, it is intended to remain largely non-technical. Graphs and tables are also employed to help you see the story more clearly. Nonetheless, you should not need to have taken (or perhaps slept through) an economics class to understand the ideas in this book. Any analytic principle that you need is introduced just when you need it, and it is kept as simple as possible. Much of economics makes sense when it is applied to a real world situation. That is the only way it is used here.

CHAPTER 2

Oil Is Cheap

There is no doubt that oil is cheap. This must be understood or all that comes hereafter will not seem worthwhile. How do you know that oil is cheap? Look around you. The stuff is used, perhaps abused, as something that is inexpensive and can be thrown away. Huge vehicles that get poor gas mileage, and a lot of them, are a clear indication that gasoline (a derivative of oil) is not yet at a price where people are concerned enough to conserve it. Petrochemicals (chemicals derived from oil) are everywhere. Plastics, pharmaceuticals, pesticides, fertilizers, and of course, lubricants are just some of the uses of oil. How did this happen? How did oil get so deeply involved in our lives? Our economy? The simple answer is that oil is cheap. When something is cheap, you use it, and not sparingly.

This chapter is about oil, energy and their economic relationship. We will first establish the relatively constant mean oil price, and provide you a brief oil history. The various energy sources other than oil will be explored so that you have a better sense of how markets react to changes in the price of oil. By the same token, energy uses need to be explored because they are also part of the market reactions. Environmental issues are intimately tied to energy use in the world, so it makes sense to discuss these issues in the context of energy use. You'll need this chapter to have a framework to understand the larger picture of energy prices and how they affect you and the economy. Remember that energy prices affect you and businesses alike. To peer into the future and have a better sense of possible investment strategies, you need a wide array of information about energy and its impact on the economy.

Oil prices will likely rise and then be pushed back down by a variety of economic forces. You should not expect the price of oil or its derivatives (like gasoline and heating oil) to move smoothly over the next 100 years. It will tend to move in fits, and then seem stable for awhile. This happens because as the price of oil rises, other alternatives become viable and oil users find the means to conserve it. In the early stages of oil price rises, other means of obtaining oil become economic. Recently, oil from sandy deposits became practical. Sometime in the near future, extracting oil from shale may become profitable. But additives are also practical. Ethanol is the most obvious, and it is already being added to gasoline, though the process is currently being subsidized. Other additives may well be introduced and utilized as oil prices rise. So as oil prices increase, something will be introduced to the market that can be used as an additive (or substitute) to slow oil price increases or erode its price level.

Oil may be cheap now, but that will not last forever. Be careful. This is not some dire warning that civilization as we know will it come to a screeching halt. Instead, it is an exploration of what happens when oil prices temporarily rise. And yes, it is possible that oil prices will rise over time. It is also possible that they will fall over time. There is no doubt that the process will be unpleasant at times. But the world is not likely to become dominated by unproductive deserts with good-looking Australians running around shooting people looking for what little fuel is left. Moreover, there is little information with which we can estimate when oil will become expensive. There are numerous theories about this, notably the "peak supply" group in the popular press, and economic models like the one suggested by Hottelling, which suggests extraction rates are related to the interest rate. The fact of the matter is that there is significant uncertainty about how much oil there is, where it is, and how difficult it will be to extract it.

2.1 A tale of three oil price histories

We've all noticed in the past 25 years that oil has become more expensive. We are all convinced that oil's price will rise, and keep rising, with no end in sight. If you look at the price of oil since 1946, you would be sure that it wouldn't be long before you can't afford to drive to work. Chances favor you would be wrong.

There are at least three ways to look at the inflation-adjusted price of oil; since 1970, since 1946 and since 1861. Your choice of time horizon very much affects what you conclude about the past and future price of oil. Of course you can fit a trendline to any data, but that does not mean that the data has a trend. Statisticians would say that the fitted trend is not statistically significant; they mean the trend can't be supported with statistical analysis. This turns out to be the case for trends fitted to oil price data. You could say, on average, the price of oil has been falling since 1970. You can also say that oil prices, on average, have been rising since 1946. Since 1861 it is not clear that oil's price has risen: it may have even declined, on average, over the entire period. None of these have statistically significant trends. Statisticians know that means are very hard to change. Trends, which are time related means, are also very hard to change. The long-term trend of the price of oil is at best a fixed mean. The data suggests to us that oil's average price is constant, and that we will likely see a return to that price fairly soon.

Consider each of these stories, from short to long term, in graphs:

The first graph, figure 2.1, is about how the inflation-adjusted price of oil has changed over the last 34 years. You probably guessed that it has risen over that period, and if we don't take into account inflation (a general increase in prices over time), that would be true. But inflation obscures the truth here. The question is how did oil's price change in terms of purchasing power. One way to think of this is: how much did a barrel of oil buy in 1970? How much did it buy in 2000? If it buys about the same amount of goods and services in both years, we could say its price did not

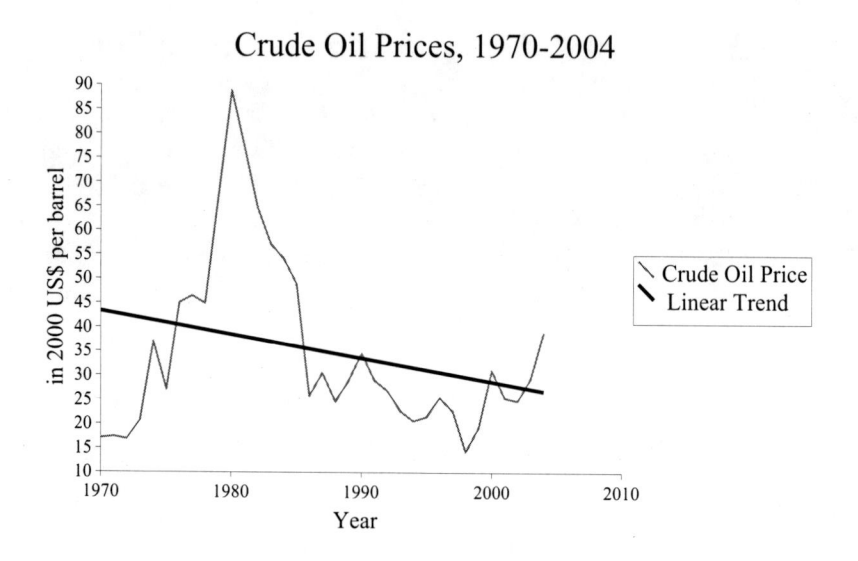

Figure 2.1: source: US Dept of Energy

change. It did buy about the same amount of stuff in 1997, and not much more by 2000. For the period as a whole, its price trend actually fell in inflation-adjusted terms. Why did this happen? You could call it the cartel effect. In the years just after 1970, OPEC (Organization of Petroleum Exporting Countries) used its cartel market power to restrict the flow of oil, and raise oil prices. It was successful for many years, peaking in 1980. But cartels are notoriously difficult to keep together. Some players began to cheat on the quotas, others, like Russia, who were never part of the cartel to begin with, used the cartel's high prices to sell oil with a nice profit. Eventually the cartel's influence weakened, and prices fell. The strong demand for oil from China helped push oil's price up in the last few years. September 11 and subsequent events have pushed it up even higher, but nowhere near the levels when the

2.1. A TALE OF THREE OIL PRICE HISTORIES

cartel was successful.

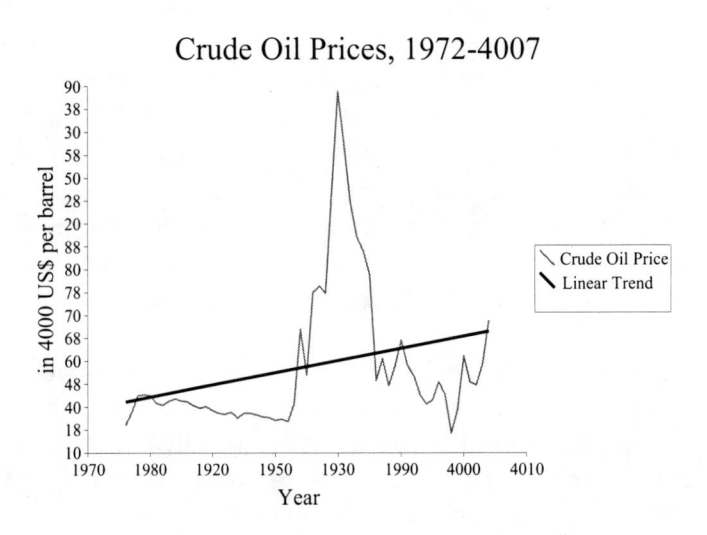

Crude Oil Prices, 1972-4007

[Figure: chart showing crude oil price and linear trend, y-axis "in 4000 US$ per barrel" with values 90, 38, 30, 58, 50, 28, 20, 88, 80, 78, 70, 68, 60, 48, 40, 18, 10; x-axis "Year" with values 1970, 1980, 1920, 1950, 1930, 1990, 4000, 4010]

Figure 2.2: source: US Dept of Energy

The trouble with time series data and analysis is that you could probably make any claim if you just select the right starting point. The claim that oil prices have declined seems to be one of those sorts of claims. Surely oil prices had been rising, and rising fast. Consider the graph for inflation-adjusted oil prices since 1946 (Figure 2.2). You should know that there is nothing special about 1946, but that happens to be when the US Department of Energy (DOE) data starts, so we'll start there. There appears to be an upward movement in the price of oil from 1946-2004, and indeed if you fit a trendline to the data, it appears that the inflation-adjusted price of oil has doubled over the period. Unfortunately, simply fitting a trendline to data is not enough. Econometric analysis of this data suggests that the trendline is not statistically significant. Economists call this a stationary process, meaning that there is one average for the data. You can say with certainty that the price of oil has become more volatile, subject to wider swings in the later part of the period. But you cannot say, with strong statistical support, that the price of oil has an upward trend. As a result, one cannot be sure that the price of oil will continue to rise.

Why does this graph show a slight rise in price, and the graph from 1970 show a decline? Once again it is the cartel effect, and the choice of starting point for the data. In the early part of the period US oil production is very strong, and much of the US energy needs are met by domestic production. An OPEC cartel would not have been very effective at this point, since the largest energy-consuming nation could meet its own needs, and ignore a cartel's market strategy. But later in the period, after 1970, the US could no longer meet its own energy needs. The OPEC cartel could apply market pressure effectively. Even though the cartel weakened by the end of the period, its effect was large over the later part of the period, and this pulled the trendline up. Given that the OPEC cartel is weak in 2005, there is no reason to believe that these high oil prices will continue. Moreover, there are many market pressures to keep oil prices low.

Maybe 1946 is not such a good year to start the analysis on oil prices. Perhaps 1861 would be better. Some the definitions of what oil barrels are change over this long period, but they have been adjusted to give a fairly consistent picture over about a century and a half. What does Figure 2.3 (on the next page) tell us about the long run inflation-adjusted price of oil? It would seem that the price of crude oil has remained about the same, perhaps even declined a bit over the last 140 years. Now that's confusing! How can this be explained? We have two important effects here: the developmental effect during the early stages of an industry and the cartel effect at the end of the period. Since we already know about the cartel effect, let's concentrate on the early period.

2.1. A TALE OF THREE OIL PRICE HISTORIES

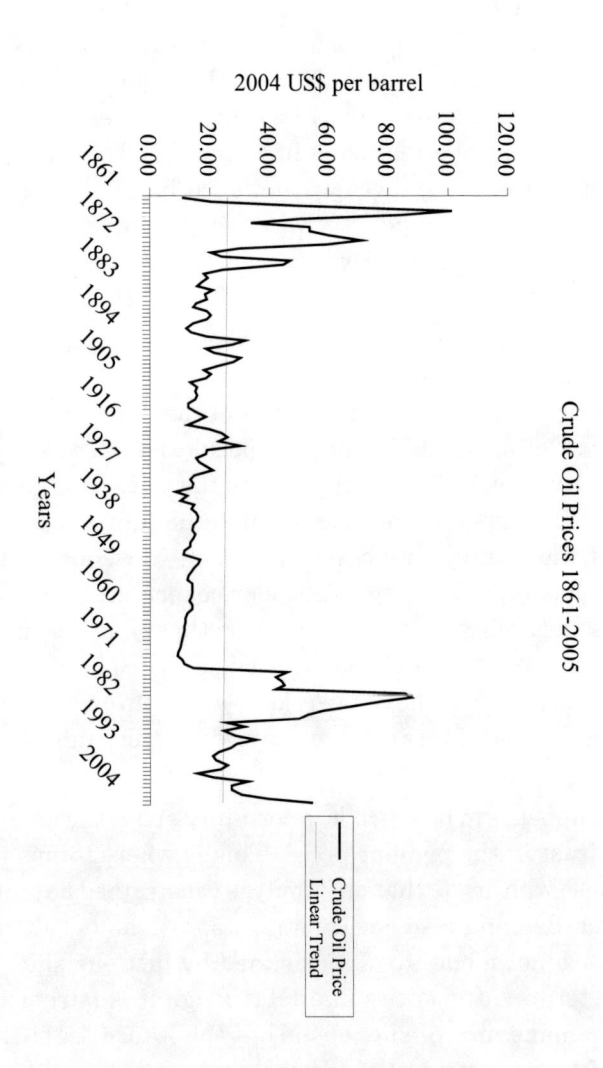

Figure 2.3: source: British Petroleum

When an industry just gets started, it tends to have a high cost structure. This is because it takes time to figure out how to produce the good or service that is being made. As the industry grows, more producers come online with new ideas in production, and perhaps of equal importance, they make the industry more competitive. Those who were there at the beginning either must reduce cost to survive, or they will go out of business. These things are the hallmarks of a maturing industry. In oil's case, as we will see in the next section, there was an even bigger issue: the chief product of the early oil industry was substituted by electricity. So the early higher prices quickly came down to a level that remained fairly even until the emergence of the OPEC cartel.

There are many forces at play when considering the price of oil, and much of the rest of this chapter should help you see some of those forces. Moreover, like the rest of this book, what happens to the price of oil in the future is speculative. The graphs and analysis above tell us what happened to the price of oil in the past, not what will happen to the price of oil in the future. Of course, we infer that the future is related to the past, so we accept the past price movements as the likely future price increases (or decreases) as well as variability. The popular notion that oil prices have risen, and will continue to rise, is not supported by past data. The past tells us that oil prices have been kept down throughout the past century and a half. Market forces have done this, and not just the supply side forces.

There appears to be much less certainty about the demand side of oil, at least in the popular press. This is where things get a bit dicey. Those who argue that oil supply is constrained but fail to recognize that demand also has constraints miss half of the market. Markets are not made up of, or defined by just one side. Supply cannot determine the entire market, though it is fairly clear that those who make dire predictions about the future see it that way. Economists gave up on that idea when they gave up Say's Law (Supply creates its own demand) more than a century ago. Markets are made up of complex interactions of supply and demand. When supply is constrained in the short term, either by natural phenomenon or by producers colluding to do so, demanders often pay higher prices. Once demanders have time to adjust, they tend to substitute away from the higher priced good. Again, demand for oil cannot rise indefinitely. Long before the world runs out of oil, the price of oil will be high enough to force its conservation. This alone tells us that demand cannot be considered constant, or forever growing, for that matter.

2.2 How oil got started

How did we get here? Oil seems so infused in our lives that we can't live without it. None of us know a time when oil was not plentiful. Its uses are found in a wide variety of common applications. It turns out that petroleum oil is a fairly recent marketable commodity. While there were minor uses for petroleum oil over the past 2000 years, the real beginning appears to be in the mid-nineteenth century. Oil's first commercial application was in the form of kerosene, and was used as lamp oil. It is notable that before this application, lamp oil typically came from whale oil, which no doubt was creating an environmental impact on at least the whale population.

The early oil industry, which was almost entirely an American endeavor, was structured to process and deliver kerosene for oil lamps. The borders of the state of Pennsylvania also geographically defined the industry. This was the period that saw the rise of John D Rockefeller, and Standard Oil. The industry grew rapidly, and enjoyed early profitability until Edison invented the light bulb (and the means to deliver electricity so the bulb could be lit). A fairly young industry at that time, oil went through a strong recession, as its primary market was supplanted by electricity. As the nineteenth century came to a close, it would have been difficult to tell if oil would be a major industry, but something important was about to happen.

The introduction of the internal combustion engine and mass-produced automobiles would change the oil industry. Again, these complementary industries were largely an American innovation, and the oil industry went with it. Texas became a significant source for oil, and due to the plenitude of oil, oil remained cheap.

Even as the oil industry expanded at a very rapid rate, the industry was not seen as strategic. It took WWII to change that. It became clear that access to petroleum was a significant advantage to the war effort, so Americans sought additional sources of oil abroad. The US negotiated with Saudi Arabia and much of the Persian Gulf in order to establish a stable supply of oil. It should be noted that this was done not due to need for new supplies, but rather to insure that others would not have access to oil. By the 1970's, however, as US reserves began to decline, foreign suppliers of oil became very important.

The peak supply theorists like to point to the decline in US production in the early 1970's as their defining moment. M. King Hubbert had suggested in 1949 that oil output in the United States would peak somewhere around the early 70's. The analysis was, all things considered, fairly good, and it turned out to be fairly accurate as well. US oil output did indeed peak in the early 1970's. Of course, Alaska's newfound reserves had not yet been tapped, and its output was diminished by a strong environmental argument. Nonetheless, the fact remains that the US peak-supply did have a peak, and this peak was forecasted correctly. One data point, or more accurately, one correct forecast does not turn a theory into fact. The current claim by peak supply adherents is that world supply is about to, or already has, peaked. It is certainly possible, but the world is a very big place, with much of it left to be explored. If oil reaches a sustained higher price, as it is expected to do, oil exploration will be of a higher priority. It is quite possible that we have found all the oil in the world. It is also quite possible that we have not.

To understand extraction rates, one can apply a simplified form of a model, known as the Hotelling Rule. The Harold Hotelling (1931) model of extraction rates suggests that any exhaustible resource, including oil of course, has an optimal extraction rate that is equal to the rate of interest. While the simplified model provides a simple rule for extraction, it is clear that in reality higher oil prices spur higher extraction rates. Higher prices also make exploration more attractive, so total extraction worldwide might rise as a result of new reserves being found, even if the rate of

2.2. HOW OIL GOT STARTED

extraction were to fall.

How much oil is extracted, and for that matter how much is searched for, at any one time is dependent on the price of oil. Oil gluts occur when too much extraction capacity is built into production to support the existing demand. This is often preceded by a relatively high price for oil. Oil shortages are the reverse: too little capacity for the existing demand. This is often preceded by a relatively low price for oil. Oil, like many other commodities, has experienced both of these market conditions multiple times. During 2005-2006, oil is in a shortage period, caused in part by unanticipated demand for oil in the world, and in part by events that limit supply. Both the US and Chinese demand for oil were pulling prices up, as were the Iraq war and a strong hurricane season, which limited supply; these forces worked together to take prices up sharply. This is likely to induce more extraction capacity and within a few years we may well feel a glut of oil on world markets. This cycle is going to continue, though if supplies actually are declining, each cycle would pull the average price of oil up over time.

The oil industry experienced several booms and busts over the past century. Booms triggered by high relative oil prices are great for investors and companies alike. These booms spill over to developers and businesses that build cities like Houston, TX. Busts usually come unannounced and drive both investors and companies into losses. That in turn spills over to developers and businesses, causing cities to empty out. That is why investing in oil companies is not always a great idea. Timing an investment becomes important, and that is not so easy to do. The 20th century shows us that the long-position investment in oil would have been a good choice, but not at all times.

The forecasts of oil prices (in 2003 dollars per barrel), shown below, provide the US Department of Energy's best guess for where the price will go over the next 20 years. They give a "reference" case - essentially the middle of their low and high case scenarios. They also give two high cases, which represent different world scenarios. There are other forecasters for this critical commodity, and some are provided here for comparison. The values seem surprisingly optimistic. Apparently, none of the forecasters believe in the peak supply theory. The increase in demand for petroleum worldwide, especially in developing countries, is likely to push oil prices higher; it is not clear that this has been incorporated in these forecasts. Also, note that these prices are in US dollars (2003 dollars to be exact), and while US inflation does not have an effect on the estimates, the US dollar exchange rate does have an effect. This will be tied together in a later chapter. For now take into consideration that a fall in the US dollar will cause oil prices in US dollars to rise.

In late 2005 the Energy Information Administration (part of the DOE) released a new set of forecasts that set oil prices in the future much higher. They now expect oil prices to be $46.90 a barrel by 2014, and rise to $54.08 by 2030. They suggest that they are now taking into account strong Asian demand (China in particular) and relatively few new oil discoveries. This would tend to keep the supply side of the market fairly tight, thus pushing up the price. It is not clear why this shift in thinking has occurred so suddenly. It appears that throughout the 1990's new reserves were discovered at a rate that was faster than the rate of extraction, and this followed a period of relatively high crude oil prices. It seems premature to ignore the economic incentives for exploration that relatively high crude oil prices create.

2.2. HOW OIL GOT STARTED

Forecasts of world oil prices, 2010-2025
(2003 dollars per barrel)

Forecast	2010	2015	2020	2025
AEO2004 (reference case)	24.53	25.44	26.41	27.4
AEO2005 Reference	25	26.75	28.5	30.31
High A world oil price	33.99	34.24	36.74	39.24
High B world oil price	37	40.67	44.33	48
October oil futures	30.99	32.33	33.67	35
Low world oil price	20.99	20.99	20.99	20.99
GII	27.08	25.58	26.66	27.12
IEA (reference scenario)	23.25	25.37	27.48	29.07
IEA (high oil price case)	37	37	37	37
Altos	21.92	22.67	23.93	24.6
PEL	25	27	27	29
PIRA	34.75	39.15	NA	NA
DB	24	24	24	24
EEA	26.58	25.55	24.93	NA
SEER	26.13	28.4	28.25	29
EVA	28.99	28.39	30.97	34.77

Table 2.1: source: Energy Information Administration, US Department of Energy
AEO is Annual Energy Outlook, GII is Global Insight, Inc., IEA is International Energy Agency, Altos is Altos Partners, PEL is Petroleum Economics, Ltd., PIRA is Petroleum Industry Research Associates, Inc., DB is Deutsche Bank AG, EEA is Energy and Environmental Analysis, Inc., SEER is Strategic Energy and Economic Research, Inc. and EVA is Energy Ventures Analysis, Inc.

Energy Prices in 2000 US$

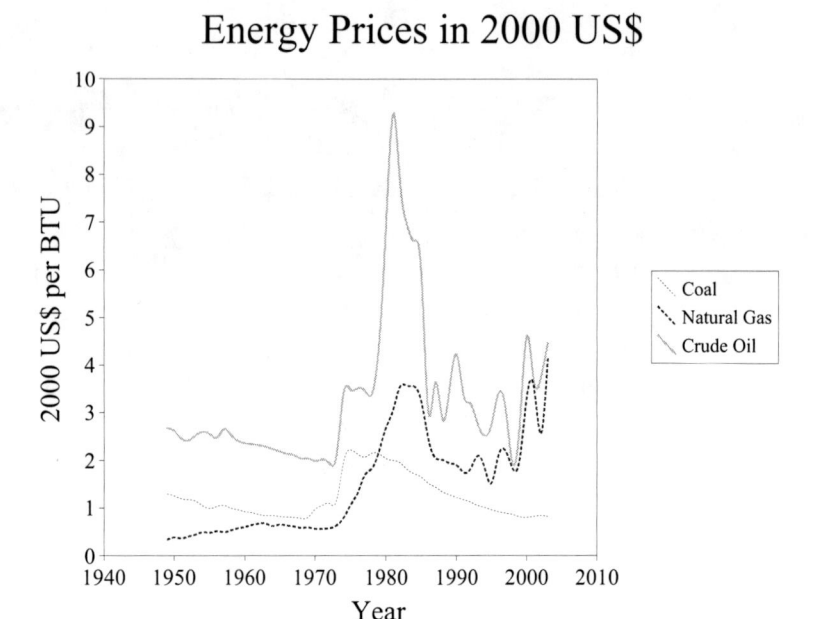

Figure 2.4: source US Dept of Energy

Of course, oil prices are not the only relevant energy price. While crude oil prices may have been relatively constant, on average, for the past 40 years, the same could be said of coal but not natural gas prices. In general, the price of coal has tended to remain about the same (on average), while the price of natural gas has tended to rise. Coal is certainly in use, and accounts for nearly half of US electricity output, but it remains plentiful relative to demand, and is further weakened by its detrimental environmental effects. Natural gas is particularly interesting because of its use as a means to produce synthetic crude (from tar sands and other fossil fuels). It is possible that the price of natural gas reflects transportation problems, so rather than strictly being a supply issue, the rise in price of natural gas may be related to its production

2.2. HOW OIL GOT STARTED

and storage.

It would also be helpful to know how the world uses energy. Over the past 20 or so years the world's demand for energy has risen by 2% per year, on average. This is consistent with the growth rate of gross world output of about 2% per year, on average. While North America has had a small increase in demand for energy (1% per year), the most rapid increases have occurred in Asia and Oceania (4% per year) and the Middle East (5% per year). Europe's energy demand grew at 1% per year, while that of Russia and Eastern Europe did not grow at all (their energy demand grew at 0% per year, on average). Both Latin America and Africa have small relative demands for energy, but both experienced energy demand growth of 3% per year.

These differences between regions' energy demand growth can be explained. First, both Europe and North America have relatively large demand for energy, so price spikes in energy prices are felt more keenly. However, because they are the most developed regions of the world, they are most able to adjust to higher prices (even if they are only temporary) by conservation and the use of alternative energy sources. Eastern Europe and Russia have not had much economic growth over this period (their economies looked like they were in constant recession), so they had little to no growth in energy demand. The fastest growing areas of the world, Asia, Africa and Latin America also had energy demand grow at about the same rate as their economies grew. The Middle East had rapid energy growth, but remember that they are major energy suppliers, so price spikes in energy prices benefit them and make their economies grow as well.

There are many names for the last century - the atomic age, the computer age, or more generically the age of technology. All of those fit, but in many ways, the 20th century was the age of oil. The industry effectively becomes a major commodity at the beginning of the century, creates many allied industries throughout the century, and becomes worrisome by the end. The seeds of energy alternatives have already been sown by the start of the 21st century. It seems reasonable to believe that this century will ultimately belong to some other energy source.

2.3 Various forms of energy

2.3.1 Petroleum alternatives you should know

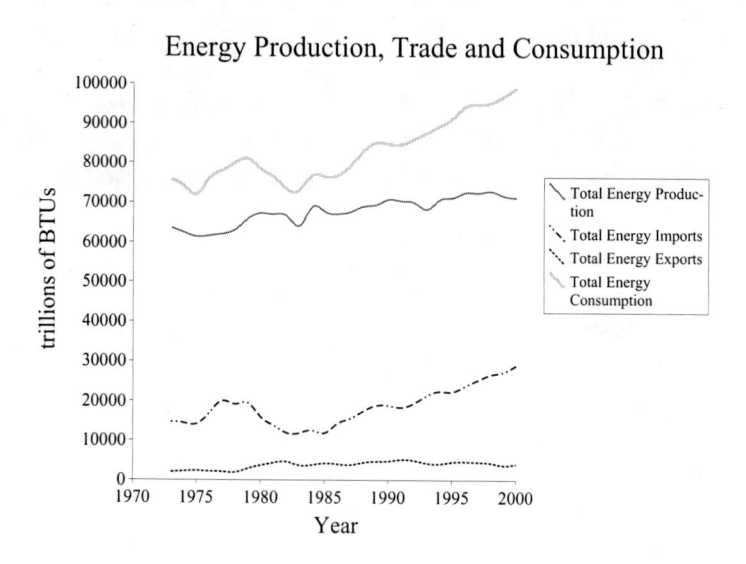

Figure 2.5: source US Dept of Energy

Heavy oil?

When we hear about oil, the discussion centers on a very specific petroleum resource: light sweet crude. Light sweet crude oil has nice properties. Among other things, it is the easiest to refine into gasoline, kerosene and other common uses. It has a relatively low sulfur content, and thus has fewer environmentally unpleasant side effects that result from the refining process. Light sweet crude is not the only type of oil, nor the only type of petroleum resource. Petroleum resources that are not light sweet crude are referred to as non-traditional reserves.

Perhaps the closest substitute for light sweet crude is known as heavy oil. This is a thicker hydrocarbon, and tends to have a higher sulfur content, as well as other impurities. Its common use is currently as an ingredient in asphalt. Given current technology, it is more difficult to refine. The refined form of heavy oil, for use

2.3. VARIOUS FORMS OF ENERGY

as light sweet crude substitute, is called synthetic crude. Heavy oil currently sells at around a 50% discount from light sweet crude. However, there may be a fairly large amount of it in the world, particularly in Venezuela. For Venezuela alone there are about 1.9 billion barrels of heavy crude. The estimated total world reserves of heavy crude are about 45.575 billion barrels, though the World Energy Council suggests that "probable reserves" of all heavy oils and bitumen (which includes oil sands) is about 193 billion barrels. If the world supply of light sweet crude oil becomes low enough, and the price high enough, one should expect heavy oil to become a common commodity alternative. This also means that refineries will need to be redesigned to manage this resource as well. This conversion may indicate a period of increased capital costs for oil companies, so it is not clear that they will remain highly profitable.

There is also another type of petroleum resource in fairly large reserves: tar sands or bitumen. The world's largest single source of this is in Alberta, Canada, with about 175 billion barrels of oil equivalence recoverable with current technology. Again, this is less pleasant than light sweet crude, and like heavy oil, more difficult to refine (and thus more costly). It is also first refined into synthetic light sweet crude. One problem for this process is that it uses natural gas as a heat source for the first stage of refining. So the costs and availability of natural gas need to be considered in extracting this alternative. But again, if oil becomes expensive enough, this is a likely alternative commodity. Current cost estimates suggest that it costs around $15 to extract a barrel of oil from tar sands, and this becomes competitive when light sweet crude is around $30 a barrel. Should tar sands become an important source for synthetic crude, there may be consequences for the price of natural gas. Unless some alternative source of fuel is found for generating the necessary heat to process tar sands, increased production will mean higher demand for natural gas. That should result in an increased price of natural gas, which reduces profitability of tar sands and makes natural gas more expensive for alternative uses.

Coal gasification, which converts coal into a form of synthetic crude, is currently in use in South Africa. It is a process that was developed in Germany in the 1920's, and helped fuel the Nazi war machine of WWII as they had little access to crude oil supplies. The lack of access to crude oil imports also pushed South Africa to develop this technology in the later part of the 20th century. The technology is competitive when crude oil prices are in the range of $30-35 a barrel. As of 2006, China is investing in coal to crude, and the United States is considering it. Besides the need for a heat source to make the process work, it appears to have another serious drawback; the process creates a significant amount of carbon dioxide.

The last possibility for petroleum extraction is shale. Since this is in the form of rock, and requires extreme heat to extract, it seems unlikely that it will become a candidate for oil use in the near future. A very new technology would be needed to make it economically viable, and while one should not count this out in the future, at this time shale should be discounted as a major resource to replace oil. However, it may be a significant resource to replace coal. It has been used as a coal equivalent for many years, and has similar burning properties. There are also very large known deposits of shale, especially in the western United States, in and around the Colorado Rockies. That makes it a particularly attractive resource for the needs of the USA. However, if natural gas is used for a heat source to extract oil from shale, the same problem exists with shale processing as with tar sand processing.

2.3.2 Some non-petroleum alternatives to consider

Alternative sources of energy

Coal has been used as an energy source for around a millenium. It continues to be used today, accounting for nearly half of all electricity needs of the USA in 2005. Its chief use is heating. In the case of electricity it is used to heat water to drive steam-powered generators. It can also be converted into gasoline, diesel and coal gas for a variety of purposes. It has not been used in converted form largely because of the availability of low-cost alternatives, namely oil. Atomic energy has been around for about 50 years. It is also used to heat water to drive steam-powered gener-

2.3. VARIOUS FORMS OF ENERGY

US Energy Supply, 2004

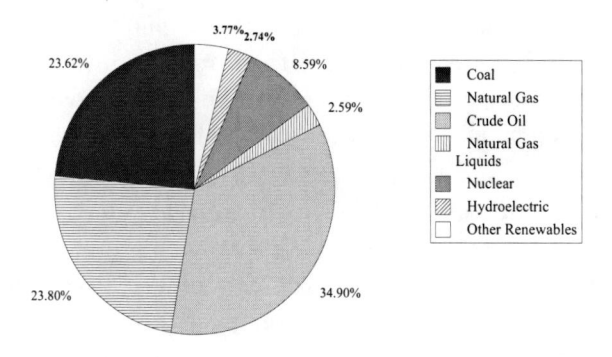

Figure 2.6: source US Dept of Energy

ators. While it cannot be used to make gasoline, it is a significant source of energy, which can be stored in batteries or more recently, fuel cells. If fusion (another form of atomic energy) can be made to work, we may well have a cheaper source of electrical power to use to replace oil, though this is expected to be at least 50 years away.

Natural gas has also been in use, mostly as a heat source, for quite some time. However, the major energy product's prices in Figure 2.4 would suggest a problem: of the three major sources of fossil fuel energy, natural gas is the only one which appears to be rising in price over time. This may be a result of rapidly rising demand, possibly from use of natural gas to convert tar sands to synthetic crude. But there is also more to this price increase than just demand. It seems that there may be transportation problems for natural gas. Natural gas is, of course, a gas at room temperature. It is also volatile. To transport it, it needs to be compressed, and shipped to a port with the ability to take the resource in. Cities don't like these ports nearby for fear of an explosion, so these ports tend to be several miles away from the shore. Crude oil and coal are easier to transport, are not especially volatile, and require no compression. Rising demand for natural gas requires increased shipping in greater volumes, making transportation costs an important component of the price of natural gas. You should be aware that natural gas requires little to no refinement, so if refinement costs for crude oil rise, it may well be competitive with gasoline and heating oil at various times in the future.

Of course, solar power and wind power already exist and are utilized in many places. While both generate electricity with minimal maintenance costs, they both involve high up-front costs. The issue is one of payback time, or how long it takes for the these forms of power generation to make up for their original investment costs. Part of this is in the interest rate (low interest rates mean less discounting of future money) and part of this depends on the cost of electricity (effectively tied to the cost of fuel). If fuel costs rise over the century, as they are expected to, both solar and wind power generation may be economically viable and profitable.

Where is the point at which this happens? At the moment, the price of electricity generated by photovoltaic solar panels is in the range of $0.30 per kilowatt hour (kWh). In the United States, the current cost of electricity is in the range of $0.08 per kWh. The trend of solar electricity costs suggests that the cost will fall, though not as quickly as it has in the past. However, if energy prices rise over the century, it is reasonable to guess that these costs might equalize, and possibly even favor solar. Wind power is already nearly competitive; its cost is in the range of $.10 per kWh. Of course, wind power is only useful for areas with a fair amount of wind. These technologies may already make sense in isolated areas without access to grid-based electrical power, but for the market to become viable, either the costs of solar power generation must fall further, or the cost of energy must rise.

There are other types of solar power uses besides photovoltaics. There are solar ovens and water heaters, as well as other solar powered appliances. It is unclear if these items will ever reach a large market. There are some countries where solar water heaters are common, and there are certainly some states where this would

2.3. VARIOUS FORMS OF ENERGY

work. Solar ovens, while remarkably cheap, are fairly rustic and are typically used by only the most committed environmentalists. Photovoltaics appear to be most marketable, and require minimum change from our current lifestyle. Moreover, using solar power should have minimal environmental impact if we use existing rooftops of homes and businesses. If we use vast solar arrays in undeveloped land, the impact is unknown. Still, if energy prices rise, one would expect other uses for solar power to become more common.

The US Department of Energy issues forecasts for energy demand and supply, including the source of supplies. The DOE is forecasting some obvious things: US production of oil and natural gas will decline, and combined with an increased demand, they expect more imports of oil and natural gas. They forecast that US energy demand will increase over the next 25 years in every category. Most of their forecasts are linear (the future values are a straight line from recent values) and are fairly simplistic for most time series under consideration. As a result these forecasts miss the clear increase in these commodities' price volatility. You should expect considerable volatility in most energy prices, even in short time periods.

It is expected that renewable energy production will increase in the next 20 years. The structure of this growth is important, particularly from an investor's point of view. The DOE expects there to be considerable growth from Biomass resources, as well as wind, and to a lesser extent geothermal. Solar power, while growing, remains a very small segment of production, and the DOE notes that the costs of solar will remain high relative to other energy forms. These forecasts suggest that an investment portfolio should include some holdings in these industries.

Coal and natural gas appear to be the best candidates for the majority of productive capacity in the near-term future. Nuclear power will remain important, and the DOE notes that nearly all of the US nuclear power generation plants in operation as of 2005 will be decommissioned by 2025. Since nuclear power has been economically viable, new plants are likely to replace them. There are two issues here: the decommissioning costs of many plants are unknown and the capital cost of new plants is going to be very large. How this plays out over the next 20 years should shed light on the next cycle of decommissioning and capitalization in the latter part of this century.

Fuel cells are currently too expensive to be part of any oil-replacement strategy. However, they are currently being developed for use as a portable electrical power source. Portable electric power can be quite useful, if you have a laptop computer and you are in an isolated area without access to electricity. Solar photovoltaics have also been used for a similar purpose. It is clear that both of these energy sources may hold possibilities for the future.

Of course, there is ethanol, an alcohol based fuel source that is currently in production. The federal government is currently subsidizing this fuel, though mostly as a price support for corn. It has been estimated that the cost of producing ethanol from corn is currently about $1.10 a gallon. Ethanol does not contain as much energy as gasoline, so we use a factor of 1.5 to approximate the energy equivalent of ethanol to gasoline. That makes the effective cost of ethanol about $1.65 a gallon. That is a wholesale price (or the price to gas stations) so if we factor in a profit margin for local retail, and taxes, ethanol becomes competitive when gasoline is selling for around $2 to $2.20 a gallon. This has already happened several times during the first few years of this century, and it is reasonable to expect ethanol to become competitive within the first quarter of the 21st century. This should make firms that are engaged in the production of ethanol more profitable, and there should be upward pressure on corn and other agricultural products as well. Clearly, rising oil prices do not damage everyone.

Biodiesel is another alternative, though only for some engines. The cost of production depends very much on what is being used as a source product. The common source material is vegetable oil (like corn or soybean oil), and is thus renewable. It is currently more expensive than petroleum based diesel, around $2-$3.00 a gallon, and is available mostly in a 20% biodiesel to 80% petroleum based diesel blend. It is clear that the environmental movement is backing this, as it has a smaller impact on air quality than more conventional fuels. Apparently, you can get instructions on how to

2.3. VARIOUS FORMS OF ENERGY

make it yourself, even out of used cooking oil, though the resulting fuel does not provide a particularly high performance.

Geothermal is an energy source that takes advantage of the deep underground temperature of the earth. In the near-ground area of the Earth's crust, say within about 100 feet of the surface, the ground temperature does not vary much after the first foot, and the ground temperature can provide heat in the winter, and a place to put heat in the summer. This energy alternative is already in use in geothermal heating and cooling systems for homes and businesses. However, this does not so much generate energy as use it efficiently.

Geothermal energy also exists as an energy source. The best known is the geothermal power plants in Northern California that use the natural geysers for steam to drive generators. These plants are known as steam plants, and can generate electricity at a cost of about 4-6 cents per kWh. As a significant energy alternative, this approach is largely limited by the number of locations where access to Earth's steam vents exist. It is likely that this can be man-made if necessary, and high energy prices may make that a necessity.

More common but less known are the binary plants that use lower temperature water in conjunction with some hydrocarbon fluids to lower the boiling point. These can generate electricity at 5-8 cents per kWh. Because these plants do not require the high heat that the steam plants do, they are easier to locate, and there are many more places they can be used. Because the hydrocarbons are not burned, but rather are recycled into the fluid, these plants are not as dependent on low cost fuel sources as coal and diesel generator plants are. That would suggest that as energy prices rise, these (and the steam geothermal plants) are likely to become more common.

Both types of geothermal plants are clearly already economically viable. A good investment portfolio that includes energy assets should also include these. The chief reason that these power plants are not more common is that they are marketed as environmentally safe. There will be more on environment issues shortly, but markets have made it clear that environmental issues are not as important as economic (profitability) issues are. For some investments that lack of foresight will be moot - geothermal energy will be profitable enough to attract considerable investment.

It should also be pointed out that hydroelectric power plants are common, and produce low cost electricity. While it is doubtful that we have reached a limit on how many of these plants can be operated, we now know that the local environmental impact can be quite large. Hydroelectric power plants rely on the holding of a very large amount of water. Hoover Dam on the Nevada-Arizona border is a good example. The dam created Lake Mead, one of the largest man-made lakes in the world. It also altered the flow of the Colorado river, which provided necessary water flows downstream. Both the upstream and downstream environments were altered, and the dam generally damaged existing fish and wildlife populations. Hydroelectric power does create relatively cheap electricity with near zero emissions, but it does have environmental consequences. It also requires the right geography with a flowing river to be effective.

2.4 The effect energy has on other markets

2.4.1 High oil prices make oil substitutes viable

To understand this, you need a little more than a rudimentary understanding of supply and demand. It is certainly true that supply and demand for oil drives much of its price in the short term. But to understand the longer-range price movements of oil, the substitutes and complements to this commodity are the keys to seeing the future price movements. Economists call these the cross-price relationships. Oil has substitutes. It also has complements. Some of oil's substitutes are fairly obvious - natural gas for instance. Small cars with high gas mileage are also economic substitutes (as the price of gasoline rises, the demand for small cars rise). Common complements would include large cars, heating homes with oil, and other uses for oil when its price is low.

Oil is also one of the most versatile commodity inputs in industrial use. You could say it has production substitutes and comple-

2.4. THE EFFECT ENERGY HAS ON OTHER MARKETS

US Energy Demand, 2004

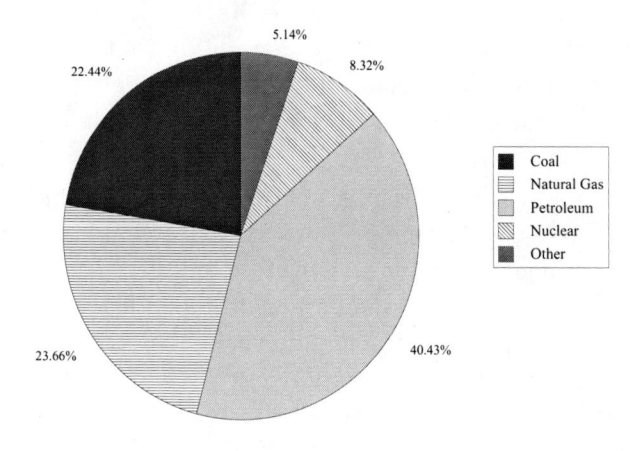

- 5.14%
- 8.32%
- 22.44%
- 23.66%
- 40.43%

Coal
Natural Gas
Petroleum
Nuclear
Other

Figure 2.7: source US Dept of Energy

ments as well. Crude oil can be refined in many ways - the most apparent are the various forms of fuel: gasoline, diesel, jet fuel and heating oil. It also is the major input for plastics. That is one reason why the price of gasoline often moves without a corresponding movement in the price of crude. A cold winter tends to drive up the demand for heating oil - more heating oil being refined means less gasoline, so gasoline inventories decline, and the price of gasoline rises. Among the most refined fuel from crude is jet fuel, and it tends to be quite volatile in its price, mostly due to the production substitutes issue.

Oil (or more specifically gasoline and diesel) are also integral to the transportation of goods throughout the world. An increase in the price of fuels drives up the cost of transportation for all goods. So there is a powerful ripple effect from fuel prices to goods prices. The 1970's saw a sharp increase in the price of oil and thus the price of fuels. The increase in cost of transportation in part pushed prices of most goods higher, and gave inflation a big push all over the world. Inflation, or a general increase in the price of goods and services over time, was ultimately sustained by monetary increases intended to buffer the shock of increased prices. Economists now have a better understanding of the nature of inflation, and can probably contain it faster than before. Nonetheless, sharp increases in oil prices will probably result in inflationary pressures in the future.

Of course, substitutes become meaningful as the price of oil rises. The most common of these substitutes already exists. For gasoline, a good economic substitute is smaller, more fuel-efficient cars. For heating, added insulation, or high efficiency HVAC. Concern about emissions is the impetus behind electric or hybrid cars, though fuel-efficiency is also part of the story. As oil prices rise, cars that use natural gas, fuel cells, and perhaps technology not yet invented will become practical. For plastics, the most likely substitute would be recycling. This is all likely to happen fairly soon, so we should see markets adjusting rapidly in the next few decades.

Nuclear fusion is probably at least 50 years away, though it may be longer than that before it becomes practical. An experimental fusion reactor is currently scheduled to be built in France. The usual assumption about fusion is that it is a clean form of nuclear power. This is likely to be incorrect. Many energy sources were touted as clean before or just after they were introduced, and turned out not to be. Moreover, a nuclear fusion reactor is a near perfect target for terrorists. So, even if we make the assumption that it is a clean energy source, there are those who will try to turn it into a risky technology.

It is estimated, based on current known oil reserves and consumption rates, that the world will run out of light sweet crude oil in 20-30 years. Of course this will not happen, as it is natural that as the supply of oil declines in the face of high demand, the price of oil will rise. This tends to diminish the consumption level of oil, and the higher the price rises, the quantity of oil demanded declines as consumers shift to alternative energy sources. You may well feel that this is little consolation, as a world beset with high oil prices is going to be difficult to live with. However, this is not

2.4. THE EFFECT ENERGY HAS ON OTHER MARKETS

the whole story.

High oil prices do diminish the quantity of demanded oil, but sustained high prices or repeated price spikes will probably cause the demand for oil to decline overall. There is nothing surprising in this, as the key to understanding this involves substitutes for oil (or gasoline). We already live in a world with oil substitutes, and these will gradually become more important as the century progresses. The most obvious substitute is natural gas, which often exists near oil fields, but has not yet been tapped. We also use ethanol as an additive to gasoline, which reduces the amount of gasoline used. Anyone who lived through the 1970's in America also knows that smaller, more fuel-efficient cars are a substitute for high gasoline prices. These are the most likely first utilized substitutes for higher oil prices. By contrast, larger vehicles, for example SUVs, tend to be complements to gasoline, and the relationship is fairly strong. It took only months during early 2004 for large vehicle demand to fall while gasoline prices rose. You should expect these substitutes to become important within the decade, and they are good areas to make investments in for at least the next 20 years. Obviously, the complementary goods to high oil prices are not good places to invest in the coming years. Remember, it is best to spread out your risk in any investment strategy, so an energy related mutual fund, or your own fund that has at least 20 different energy stocks, would be a better choice than one stock or a few stocks.

The goals of reducing greenhouse gas emissions and the need to provide energy to the world would seem to be at odds. Under current technology these goals are not mutually consistent with each other. But there should be technologies that can achieve both goals simultaneously. Nuclear fission does appear to meet both goals, but the world has largely rejected this technology as being too expensive and risky. Nuclear fusion holds much promise as an energy source that can achieve both objectives, though it is still a long way off, and may have other unforeseen problems. The desire to develop energy technology to solve this problem is fairly high, though there is not much in the current set of opportunities that warrants serious investment.

One novel idea is to use methane emissions recovered from landfills or industrial processes, refine it and use it as an energy source. The concept is appealing because it reduces greenhouse emissions of methane, and utilizes it in more economic way. Presumably, the burning of methane will ultimately contribute to greenhouse gases, but at least the net effect will be nil, compared to not having recovered it in the first place. Of course those advocating this approach suggest it will reduce greenhouse gas emissions by some specified amount, when in fact it is not likely to. If it truly substituted some other fuel (like oil) then it would have that effect. What is more likely is that it will become another energy source for a world with an ever-increasing energy demand.

That leaves only one major avenue to respond to higher energy costs: improve efficiency. Automobiles that get higher gas mileage, either by making them smaller or by increasing engine efficiency, is one way. Using more insulation and reducing air leaks in buildings is another. Homes and businesses can use more fluorescent lighting, and can cut the use of lighting and heating or air conditioning during non-use periods. Geo-thermal heating and air conditioning is also a good candidate for improved efficiency in energy use. There is a limit to how far this can take us in reducing energy use, but we are not yet close to that limit. You should reasonably expect companies that are involved in these types of efficiency improvements to have a growing demand for much of the century.

By 2030 these obvious substitutes may well have peaked. This leaves the energy alternatives we have now, but are in early stages of development of application. Solar power, fuel cells and emerging energy technologies should bear fruit by 2030. These alternatives should become more important over time, but reach prominence only once the existing technologies have run their course. As energy prices occasionally spike over the next two or three decades, more research funds will be drawn into these alternatives, and we should expect to see some breakthroughs in these areas.

Prudent investing in existing alternative energy technologies should be a good choice over the next 20-30 years. There are many segments of the alternative energy industry. Wind power and solar power are both emerging as viable energy sources. In the short term, we ought to expect these to be the most likely success stories. Natural gas and the applications of natural gas should also be good investments in the short term.

In the longer range, fuel cells are as likely as any other technology to be successful. It has a major advantage over other fuel-burning technologies: its chief by-product is water. This technology already has been applied to small-scale power needs, and will likely be developed for larger scale applications. It does not seem likely to be the economic choice for what might be called wide-scale applications like power generation for cities. But it might be economic for use in automobiles and other places where portable power is desirable.

Nuclear power, while already in use, appears to have reached a level of use where it is not likely to go much further. The most important reason is that it has not been highly profitable. In some cases it was applied where or when the demand for electricity was not large enough to support the level of output that made it profitable to run the reactors. Government regulation, clearly necessary for the industry, also made it difficult to maintain profits. But there is also considerable political resistance to nuclear power, so the future of the nuclear (at least fission) industry is not particularly bright.

2.5 High energy prices hurt some industries

2.5.1 The damaging effects of high oil prices

Energy has complements. Economists call complements those goods or services whose demand is inversely related to another good's price. These tend to be goods that are used with the good in question. That is a bit looser than the demand concept, but for our purposes, it is close enough. For energy, goods that use a lot of energy would be complements of energy.

Oil, or more specifically, gasoline, has a strong complement in large cars, or cars that get poor gas mileage. When fuel prices are low people are willing to use it inefficiently. This is why the demand for large cars and SUVs has been high during periods of relatively low gasoline prices, and why these cars become harder to sell when gasoline prices rise.

Oil and other petroleum products have a wide array of strong complements. Heating oil and furnaces that use heating oil are complements. Propane and furnaces that use it are complements. Insecticides, pesticides and fertilizer are also complements. Plastics are complements. And, of course, most lubricants are too.

Allied industries also are effectively complements, though weaker than those above. Air travel is a complementary industry, though jet fuel is only one component of the cost of running an airline, it is certainly a major constituent cost. Since most trains and ships use diesel, they too are complementary, but they tend to be much more efficient in their use of fuel, so the relationship is weaker. Since lodging (hotels, motels, etc) and tourism are dependent on travel, and that is dependent on the cost of travel, these are also complementary industries.

To illustrate these relationships, consider the effect that hurricane Katrina (2005) has had on energy related industries in the US economy. The hurricane damaged both the crude oil and the refining infrastructure. The damage was fairly large. Something on the order of 15% of refining capacity was halted, at least for a brief time. Airlines were uncertain of the availability of jet fuel, and maintaining adequate fuel supplies as well as paying higher prices was a major worry. For each of us, the price of gasoline spiked. Some of us cut all unnecessary travel, including the Labor Day weekend. This had an unpleasant impact on the tourism, lodging and amusement parks industries. The cost of transporting food, especially perishables, rose; and this translated to higher prices in the grocery stores. Once things settled down and production resumed, prices and activity returned to normal - but allied industries were still affected by the lost business during the period.

2.5.2 High energy prices have damaging effects too

Oil and other petroleum resources are not the only sources for energy, so we need to think more generally about industries that are

2.5. HIGH ENERGY PRICES HURT SOME INDUSTRIES 35

related to energy as complements, and chiefly as production complements. These are industries whose products are heavy energy users. The lighting industry comes to mind, though we should be careful to clarify this to exclude specialty low energy bulbs like compact florescence that act more like energy substitutes. Incandescent bulbs are clearly energy hungry, and meet the definition of a complement.

CRT picture tubes (used in most televisions and computer monitors) are also heavy energy users, though LCD monitors are less so. A typical household oven and a clothing dryer are among the most energy intensive appliances in a house. Furnaces use a large amount of energy, though this is spread out across the energy industry because there are so many different sources of energy that are used (for example, heating oil, electricity and natural gas). Air conditioning is a very heavy user of electricity.

The home uses of energy, at an individual level, are quite small when considering industrial uses. In aggregate, home use of energy is a substantial portion of total energy use - but this is likely to change much more slowly than industrial use, because the incentives to individuals to alter energy consumption are not as great.

It is clear that industrial activity is a supply complement to energy markets. So as energy prices rise, we should expect industrial activity to decline (or substitute energy use, perhaps by conservation efforts or changing the type of energy used). For the United States, this is particularly interesting because much of the heavy energy uses in industry are disappearing from the American industrial structure. The US economy has been changing in response to free trade, and that change has been the loss of many heavy industries, with higher tech industries replacing them. In many ways, US industry is in a better position to deal with higher energy prices than ever before. Should the US take the lead in alternate energy production (for example, wind, solar or nuclear fusion), it would be in an excellent market position as a world energy supplier. If some other country or countries take the lead, the US may have missed an opportunity to create (and sell) valuable resources, but it should still be in fairly good position in world markets.

While the next section deals with substitutes, it should be mentioned here that as any energy price rises, the heavy users of that energy source will tend to try to substitute away from it. So if oil prices rise, firms will try to use more natural gas as a heat source, or some other form of energy. However, it is expected that energy prices will repeatedly spike over the century, so there is a limit to how far firms can go. Eventually, high energy using firms, no matter what form they use it in, will experience rising costs - this will force their output prices up and probably squeeze their profits as well during oil price spike periods. Firms that are not heavy users of energy should be able to maintain their footing in such a world, and should be able to buffer the variation in energy prices.

2.6 Environmental issues to consider

The environment is clearly tied together with energy issues. Both air and water pollution are part of this story, and since the Clean Air Act of 1963 (with revisions in 1970 and 1990), the United States has shown a willingness to make adjustments to output for the improvement of the environment. You might note that 1970 also marks the creation of the Environmental Protection Agency (EPA). Many believe these efforts have not gone far enough, and they point to global warming as the chief result of our poor stewardship of the planet. There is much to support that we have not done well in taking care of our environment, but the data is far from conclusive.

One of the most interesting elements of the energy-environment connection is that virtually all of the current production technologies (fuel cells, biodiesel, solar, geothermal, etc) were developed not because the cost of oil was high and needed a replacement energy resource, but for environmental reasons. That is important because nearly all alternative energy sources are currently marketed for their lower emissions, not their lower costs of use. Sometime in this century that will probably change.

There's an interesting twist here. As petroleum based resources begin to decline, and their price naturally rises, we will be guided to replace those resources with alternatives, probably those men-

2.6. ENVIRONMENTAL ISSUES TO CONSIDER

tioned above. Even if we do nothing, market pressures will ultimately force us to conserve our resources and reduce our air pollutants or at least those from fossil fuels.

This seems to be missed by the various commentaries by scientists. It seems to be assumed that oil, natural gas and other fossil fuels will run out. This often comes from simple linear forecasts of past consumption levels, with some constant growth rate in demand. If this forecast is really true, we probably have an ecological and economic nightmare awaiting us. The amount of air and water pollution from the burning (and other uses) of fossil fuels is quite large; should we use these resources in this steadily increasing fashion there probably will be serious environmental damage, and there are indications that damage is already occurring.

But we won't use these resources that rapidly. The market allocation mechanism (price) will eventually force us to change. The real issue is whether our past use, and our consumption of petroleum in the near future will cause irreversible harm. We have some really obvious damage with oil spills and air pollution in our major cities. Beyond that it is clear that the amount of carbon dioxide in the atmosphere (a greenhouse gas) has risen over the past century. It is well documented that Earth's temperature has risen over the last century, though it is not clear that the variation is outside the normal range of variation of temperatures for the planet.

Most of the arguments about global warming trace the beginning of the problem back to the first industrial revolution (about 1770). That is probably overstating the case. The first industrial revolution saw an increase in the use of coal, and steam was the driver of locomotion in engines. The scale of use of fossil fuels was fairly small, and given the size of the planet, not significant. The better beginning might be the 2nd industrial revolution (about 1880), which was more about chemicals and electricity. Still, the start of the 20th century is where petroleum (in the form of gasoline) becomes heavily used. So the right argument would seem to start in about 1900. That implies that the bulk of the cause of global warming is related to our use of oil. There are additional chemicals that may be damaging, for example, CFCs (chlorofluorocarbons) that are believed to be depleting the ozone in our atmosphere. We have certainly used many chemicals without fully understanding their effects on our environment. Land use change (for example deforestation) is also of consequence, particularly for underdeveloped countries. We have also used a very large amount of fossil fuels.

There are some arguments about how much effect we have had, and what part of the warming is part of Earth's natural cycle, but is it clear that the global temperature has risen, and the science indicates that we are responsible for some part of that.

The science on global warming and fossil fuel use is on solid ground, and we should be concerned about it. The economics of global warming - how markets might be involved in correcting it, and the effects on economies resulting from government policy that addresses the issues are far from clear. You should be aware that no one knows exactly how global warming effects specific regions. It is entirely possible that weather in the USA will improve (become wetter) while other regions could become drier. Of course the reverse is possible as well. The world as a whole is not likely to be better off warmer, but the distribution and extent of damage is unknown for any country or region.

Global warming and fossil fuel consumption is going to be a major issue for the 21st century. Knowing that tells us to expect price variations in a variety of assets that are influenced by government policy. These assets include oil company stocks and bonds, alternative fuel assets and perhaps more significantly the stocks and bonds of companies that produce products that use fossil fuels. The market complements of fossil fuels - automotive and allied transportation industries, petrochemicals including pesticides and fertilizer, and plastics are going to feel the effects of changes in government policy. By the same token, businesses that are environmentally more appealing, like the geothermal and solar energy industries, are likely to be boosted by changes in government policy. Rather than looking for a price where this happens, you should look for shifts in government policy, likely to be well publicized. Keeping up with the news should be sufficient.

2.7 A quick comment on water issues

Water quantity and quality issues are important components of the environment. Water issues have probably been around since the dawn of civilization. The Romans did not build aqueducts two millennia ago for fun. Water was clearly important to them. In a great many parts of the world, water is a vital resource that needs constant attention. There are relatively few parts of the United States where this is true, but the Southwest and California are important areas and they have pushed the limits of prudent water use. Cheap and abundant energy can help solve water problems, but with rising energy demand, it is difficult to know when energy will become neither cheap nor abundant.

Considerable effort has gone into utilizing the Colorado River in both Arizona and California. This effort has been largely successful in creating agricultural areas in what would be otherwise desert. But it is already at the limit of use. There are few alternatives to fresh water resources, so the system is not capable of much adjustment. Given the population being supported in the desert and southern California, the system as a whole could be described as critical. It will not take much for there to be serious water problems for the entire area. It is reasonable to expect the region to be fairly risky for the next century. There may be great opportunities in real estate and natural resources, but there is also the possibility of great failures in the region as well. If the situation gets bad enough, we may even see an exodus from the region to places that are more stable in their water resources.

If there is a reversal of the population flow into California and the Southwest, there is a very real possibility of a dramatic change in real estate prices in both areas. Real estate costs are fairly high in the region, especially in California, though given the coastal terrain and the state's economic growth history, as of 2006, they are probably sustainable. However, if there were a severe drinking water crisis that remained prolonged, the real estate prices would be no longer sustainable. So it is possible that sometime during this century California and the Southwest could see a decline in real estate prices.

By the same token, an exodus from the southwestern region means that somewhere else will become a destination for the displaced residents. The Midwest seems as likely as anywhere else, and generally has good water supplies. Real estate in this region could see a sharp increase in valuation as new residents chase after a fixed amount of land. Of course, those leaving the southwestern region could end up all over the country, so the effect of their movement could be mild in real estate valuation terms.

Water resources, notably in the Midwest US, remain plentiful. The Pacific Northwest and Atlantic states are also fairly abundant in water resources. But the Plains and Southwest are very much at the whim of nature, and water problems are likely to get worse over the century. The Southwest is particularly important, as it has enjoyed the largest growth rate, and some of the highest increases in real estate value over the past half-century. An exodus from the region would mean a decline in real estate values, as well as a decline in the region's output and tax revenue. The US as a whole would not experience this, but there would be a sizable shift in wealth across regions. The winners would be in the water rich regions, the losers in water poor regions.

You might also note that ethanol, at the moment one of the highly thought of gasoline additives or substitutes, requires a significant amount of water in its production. Not only is corn grown in the Midwest, but the water resources to take corn and transform it into ethanol are also in the Midwest. While it seems unlikely that those water resources will become a problem, it is certainly possible that some localized water issues could become important, especially during drought years.

Remember that this is a long-range forecast - this seems like a possible outcome over the next 100 years. In the shorter term real estate values may well continue to rise in the Southwest, as they have for quite some time. The turnaround in land values may also result from some other cause, but it is clear that there are serious water shortages in the region, and they are not likely to get better. It only takes a single crisis, and people will get the message that they might be better off elsewhere.

2.8 Want to find out more about energy?

Information about oil, and energy in general, is both vast and deep. If you like reading, there are books and websites devoted to the topic. If you are a numbers person, data is also available on the web. If you prefer television, a recent series by PBS was done on oil. It should be easy for you to follow up on anything suggested here, and draw your own conclusions.

You should know where I drew my ideas and data from. Let's start with the oil history section. I started with a National Geographic article from June 2004, which was on 'The End of Cheap Oil', by Tim Appenzeller. I drew more from the PBS website (http://www.pbs.org/wnet/extremeoil/) associated with the series *Extreme Oil*. The site is not only rich with information on oil's history, but also links to other sites with greater detail and data. A more detailed history is available in a book by Daniel Yergin, *The Prize: The Epic Quest for Oil, Money, and Power*, from Simon and Schuster published in 1991.

I also used the US Department of Energy (DOE) as a primary data source. The Energy Information Administration (EIA), which is part of the DOE, maintains a site (http://www.eia.doe.gov) which is great place to start. It is very easy to find data, mostly in tables, and downloadable into spreadsheets. My data on the real oil price series from 1946 on comes from there. Almost all of the data used to support the graphs, including the pie charts, also come from there, the exception being the data to support the long term oil data from 1861, which I found at the British Petroleum website (http://www.bp.com) in their Annual Report. Some of my information on reserves of oil came from the World Energy Council (http://www.worldenergy.org) - check their Survey of Energy Resources for more details on both reserves and issues related to alternative fuel's usability.

To find energy alternatives, I used a search engine (I prefer Vivisimo, which clusters the results from other search engines) on a variety of keywords. The results are remarkable. You can find individuals selling all sorts of energy saving devices, from solar ovens to wind power generators to photovoltaics. But you also find a lot of information about energy alternatives. I linked to the DOE's Energy Efficiency and Renewable Energy site (http://www.eere.energy.gov/) for information about wind, solar, geothermal and biomass energy.

For more information about Hubbert's Peak (the idea of peak oil) take a look at http://www.hubbertpeak.com/ for a very one-sided view. The internet is filled with sites that seem to want to announce that the peak is either past, or is coming soon. They might be right, but every few years there are predictions that world will end as well; so far as I can tell, that hasn't happened either. Somewhat more relevant, though not as much fun, is the Hotelling model of optimal extraction rates. One of his papers that deals with this is in 'The Economics of Exhaustible Resources', *Journal of Political Economy*, 39, 1931. For a more complete view of this topic, pick up a book on Natural Resource Economics.

If global warming interests you, consider the Marian Koshland, National Academies of Science, Pew Center as a source for information and data. It can be found at http://www.koshland-science-museum.org/exhibitgcc/ and is full of great graphics and data that you can use to form your own opinion. Other environmental resources can be found at the Resources for the Future homepage, at http://www.rff.org/ which covers a wide range of topics. The National Oceanic Atmospheric Administration (NOAA) climatology page can be found at http://www.ncdc.noaa.gov/oa/climate/climateresources.html. The US Environmental Protection Agency (EPA) should also be included in these issues, and they can found at http://www.epa.gov.

CHAPTER 3

A World Divided: Trade And Finance

3.1 How international trade might change

International trade has been a topic in Economics since the field began. For much of the history of Economics, economists have been arguing, some even screaming, that free trade is preferred to all other possible trade structures. Governments and many interest groups have been arguing, some yelling, that free trade is not the best, at least not for them. It's entirely possible that both are right. Confused? Read on, by the end of this chapter, you may well see how this is possible.

We will start off with a discussion of free international trade and the state of the world trading system in 2005. Then the unpleasant side of trade will be covered, and you should be able to get a sense as to why there are those who feel free trade is not best (at least not for them). Globalization, which is international trade taken further by the use of outsourcing, insourcing and job relocation, is discussed, and the arguments against such practices are sketched out for you. I then highlight three countries that have special circumstances: China, India and Brazil. Because I argue that the world will move away from free trade, these countries have special opportunities and challenges, mostly guided by their proximity to large well-developed economies.

Trade issues are only one part of the international sector, so we then turn our attention to financial issues in the United States and the world. For the USA, large fiscal and trade deficits are seen as a problem that must be reckoned with, though there are reasons why they can be maintained for a brief time. The world has had to adjust to US financial needs, but the analysis suggests that this will not last. How the USA fixes its problems, and how the world values the US dollar are critical issues now and in the future. Because US economic growth is central to this story, I have included a discussion of what's likely for future growth.

Why is the international sector so important? Because it sets up the flow of resources and income in the world. What countries succeed and who in those countries do well are important questions to answer in any story about a future that is dynamic and variable. Understanding the international economic story about to unfold helps you include international assets in your portfolio, and also helps direct your investment choices domestically. It also helps you see a future where some countries have opportunities that simply do not exist in 2005. By the time you finish this chapter, you should gain some insight into these issues, and perhaps also gain (if that's the right word) a little fear of US fiscal and trade deficits.

3.2 Free trade is nice

The movement of the world toward free trade has been largely successful for the past 50 years. At this point (2005) there are no countries left in the world that do not engage in trade with other countries. A great many countries are now trading with few tariffs or other impediments to trade. Several prominent trading groups have emerged, notably the EU (European Union) and NAFTA (North American Free Trade Agreement). It is easy to see from Figure 3.1 that trade has grown very sharply over the last 50 years. Economic theory infers that countries gain when they trade

3.2. FREE TRADE IS NICE

with each other in most circumstances. The same theory tells us that in every country trade hurts some groups within the country.

It is hard to guess how long the current run of countries moving toward free trade will last. Most economists believe that trade is a good thing for all countries. Nonetheless, there are many groups that are hurt by trade, even if the country as a whole gains. The political pressure that is brought to bear by these groups is likely to ultimately turn the tide on those who advocate free trade. In well-developed countries, this is typically the labor-using industries. These industries include textiles, manufacturing and some services. In less developed countries these industries tend to be heavy capital users, or industries that tend to be risky, such as banking, insurance, software development and entertainment.

The obvious countries that are likely to be damaged by the shift away from free trade are the Southeast Asian group, especially China. But India and Brazil are also dependent on trade, and are likely to sustain considerable damage to their economies, should trade volumes decrease. Mexico, while also dependent on trade, is likely to escape from such a reversal, as it is both a border country to the USA, and covered by NAFTA. We are probably going to see trade break down to trading blocs, like NAFTA and the EU. Trade will continue within blocs, but inter-bloc trade will decline. Since countries tend to set up trading blocs with other similar and geographically proximate countries, this means that well-developed countries have blocs with each other, and underdeveloped countries will do the same. This is unfortunate, as the greatest gains from trade occur when dissimilar countries trade with each other.

International trade is a major factor in the economies of the world today. For the past 50 years it has also been a significant part of the world's economies. It is almost certain that it will be a driving force for this century. The fact that we see so many products from all over the world in nearly every country leads one to conclude that trade has always been important. That is not entirely true.

Trade amongst the world's economies really took off after the Second World War with the introduction of GATT (General Agreement on Tariffs and Trade). There have been successive rounds of these agreements and they have generally pushed the world toward trade with fewer restrictions. The process has been so successful that a formal organization was created to manage these agreements, the WTO (World Trade Organization). It might be possible that trade restrictions have been reduced to the point that there is very little more that can be done without unpleasant political fallout. It also appears to be the case that many of the past trade agreements tended to benefit the developed countries more than the underdeveloped ones; many underdeveloped countries are beginning to resist new agreements. The current round of these negotiations (called the Doha Round) may well be a harbinger of a more difficult future in international trade.

Figure 3.1: source IMF and US Dept. of Commerce

For the past 50 years, trade has expanded faster than production for most countries. This suggests that trade has led the world's economies. The pattern of this trade has changed considerably over the past half-century. In the years just after WWII the United States was one of the large exporters to the world, mostly in finished and capital goods. Underdeveloped countries were also significant exporters, but mostly in raw materials and foodstuffs. By the 1970's, the United States became a net importer from the

3.2. FREE TRADE IS NICE

world, and most developing countries had stagnated in their exporting. The next 30 years to the end of the century saw a dramatic shift in world export leadership. What became known as the Asian Tigers (South Korea, Singapore, Taiwan, Malaysia, and Indonesia) became prominent in exports. The nature of these exports changed as well - technology-based and finished goods were their chief exports. As of 2005, China is the largest exporter to the United States and much of the world. Canada and Mexico are the largest trade partners (ranked by total trade, exports + imports) and China is third. Given the remarkable change in both the structure of trade in the world, and the nature of goods being traded over the past 50 years, there is every reason to believe that these patterns and structures will change many times in the next century.

How these patterns change is very much a part of the story of this century. The forecast here is also the most speculative forecast in this book. There are, however, some signs of what might be coming, and it is not all bad. One signal of change in trade structure is the formation of trading blocs. Most folks have heard of NAFTA (North American Free Trade Agreement) and the EU (European Union). There are also several more that very few Americans have ever heard of. Consider Table 3, on the next page:

There are more, smaller economic cooperation pacts that exist in addition to the table above. The idea of economic integration (countries binding their economies together to make a larger market) would seem to be in its heyday. Larger markets tend to improve productive efficiencies and make more products available to the countries' consumers. It also tends to enhance the group of countries' bargaining power in trade negotiations with other countries, since a larger export market is at stake. These positives are what is pushing trade integration into a part of the story of the 21st century.

Table 3 Trade Blocs

AFTA *Association of South East Asian Nations (ASEAN) Free Trade Area* Brunei, Cambodia, Indonesia, Lao PDR, Malaysia, Myanmar, Philippines, Singapore, Thailand, Vietnam
Andean Pact Bolivia, Colombia, Ecuador, Peru, Venezuela
CEFTA - *Central European Free Trade Agreement* Bulgaria, Czech Republic, Hungary, Poland, Romania, Slovak Republic, Slovenia
ECO - *Economic Cooperation Organization* Afghanistan, Azerbaijan, Iran, Islamic Rep., Kazakhstan, Kyrgyz Republic, Pakistan, Tajikistan, Turkey, Turkmenistan, Uzbekistan
EU - *European Union* Austria, Belgium, Denmark, Finland, France, Germany, Greece, Ireland, Italy, Luxembourg, Netherlands, Portugal, Spain, Sweden, United Kingdom (added May 1, 2004) Czech Republic, Estonia, Hungary, Latvia, Lithuania, Malta, Poland, Slovakia and Slovenia
Mercosur - *roughly translated: Market of the South* Argentina, Brazil, Paraguay, Uruguay
NAFTA - *North American Free Trade Association* Canada, Mexico, United States
SADC - *Southern African Development Community* Angola, Botswana, Congo (Dem. Rep.), Lesotho, Malawi, Mauritius, Mozambique, Namibia, Seychelles, South Africa, Swaziland, Tanzania, Zambia, Zimbabwe
SAPTA - *South Asian Association for Regional Cooperation Preferential Trading Arrangement* Bangladesh, Bhutan, India, Maldives, Nepal, Pakistan, Sri Lanka

However, trade integration is not a purely positive choice for the countries involved. Because it binds a group of countries together, a weak economy in one can put a drag on the others in the group. Most economic policy is strongly linked to a country's politics, so integration means that each country in the group has a stake in each other's politics. That can lead to some uncomfortable relationships between countries within the group. There may also be pressures to switch groups, should a better offer come along. In short, there are reasons to believe that while the world is moving toward economic integration now, there may soon be a time that this will reverse.

3.3 Resistance to international trade

Reversing economic integration is not likely to be easy. This is particularly true for countries in the European Union (EU). They have had some form of a common market for the last 50 years, so there is considerable co-dependence in their economies. Most of Europe also joined in the Treaty of Maastrict, which bound them to a single currency (the euro). This makes their integration even tighter, and unraveling it even more difficult.

Because countries enjoy considerable gains by binding their economies together, there are strong economic forces keeping things that way. These forces have to confront political as well as economic forces to break with the unified markets that exist. While larger markets mean lower prices and a greater variety of goods and services, they also mean changes in the types of jobs available, and where those jobs are located within the unified market. Capital and entrepreneurship tend to be more mobile, and are more likely to benefit from the increase in opportunities that come from large markets. This creates the basis for economic tension within countries about their place in unified markets.

It can be shown that countries gain from an increase in trade due to lower trade barriers, but it is also known that the gains from trade are not equally distributed throughout a country's society. This is where politics comes in. One of government's roles is to determine how to deal with income distribution issues. Some, like the United States, tend to do less, while others, like many in Europe, tend to do more. Economic integration upsets the balance that existed before integration, so governments have to deal with issues that arise from the new structure. This often turns out to be a fiscal policy problem that spills over to the unified market as a whole. If the countries have unified their currencies (like the European Monetary Union) this expresses itself quicker. Because other members of the unified market have a stake in these decisions, they are likely to become involved in each other's politics. This creates the basis for political tension within the system.

These tensions push and pull on the original agreement that binds economies together. Over time, divisions become focused and intensified. Ultimately the countries face a decision: either full political and economic integration, or go it alone. There is no history of consensual political unification, though it is possible that something new in this realm will happen in this century. Nonetheless, it seems more likely that these economic unions will break apart.

What does disintegration look like? It is likely to be a slow process, though quicker than the time it took to integrate economies together. It took approximately 40 years from the Treaty of Rome (to create the European Common Market) to the introduction of the euro. It will probably take less than a decade for the whole thing to unravel. Political squabbling and economic dissension will result in small steps away from a unified market. Some trade barriers will show up in ways that don't look like tariffs or quotas, but have the same effect. One country will say that the food products or manufactured goods of another fail to meet some standard, and will stop importing it. The country whose imports are being turned away will do the same. The trade levels within the unified market will begin to decline. Finally, one country will place a protective tariff on goods coming from others within the unified market, and it will be over.

While this could lead to a war, the alternative of people easing off seems more likely. Politicians will lament the demise of the common market, and say it was a nice experiment. Countries will return to protected trade, though they will still trade more with those countries that were once in the unified market. They will

3.3. RESISTANCE TO INTERNATIONAL TRADE

drop the unified currency (if they have one), and return to their old currency. Their respective differences in both fiscal and monetary policy will appear quickly. Within a decade of the breakup, it will seem as though nothing had changed from the time before they were together.

Currencies have always been volatile, risky investments. Should the breakup of common markets come to pass, they will be more volatile than their historic norms. Nothing drives a currency up and down like uncertainty. From the beginning of the political squabbling, currencies will move. Rumors of a pending breakup will drive the unified market's currency or currencies down. Rumors that all is OK will drive the currency back up. There may be a few years of this, and investments with foreign assets will be riskier. An options position called a straddle could be profitable during these events. A straddle is essentially the following: both puts (short) and calls (long) are held and sold while the market dips and peaks. If you choose to do something like this, be aware that such investments are not for the timid and are extremely risky.

A breakdown in trade of the type suggested here is not intended to be dire. Right now, due to the enormous energy that was put into GATT (General Agreement on Tariffs and Trade) and its successor, the WTO (World Trade Organization), the world has moved fairly quickly toward a freer trade position. We should anticipate this reversing sometime in the near future. The combination of politics and economics is the driving force behind such a change. The problem lies in the effects of trade on both the importing and exporting countries. Consumers of the imported good (and this can include industrial inputs) are winners in the importing country, while the exporting producers in the exporting country are the winners as well. The inverse is typically true for each country: import-competing producers and the consumers of exports in the exporting country are the typical losers from trade. As a result, distinct political groups with very real economic interests appear in both countries, and eventually they make their desires known. So, even though it can be shown that every country as a whole stands to gain from trade (at least in the short term), the gains have not, in practice, been distributed evenly across the country's society, and thus the essence of anti-trade politics.

3.4 Globalization and anti-globalization

The forces of trade and anti-trade politics are currently visible in the debate over globalization. The essence of the argument for globalization, or greater openness and trade, is that as long as transportation and information costs remain relatively low, countries can specialize and export those goods and services that they are comparatively good at producing. Countries with abundant labor tend to specialize in goods that use labor in their production. China clearly fits this, and produces clothing and other textiles at low cost. The USA is abundant in capital, and tends to produce goods that use capital in production, like commercial jetliners and heavy machinery as well as financial services at low cost.

Once trade barriers like tariffs and quotas are removed, countries have a strong tendency to export their low cost goods. Why? It is simpler than you might think: the prices of the goods exported are higher in the foreign country than at home. OK, so then why do countries import goods? Again the simple answer: imported goods are cheaper than equivalent goods made at home, or there are no equivalent goods made in the home market at all. The USA imports both bananas and coffee because it lacks the conditions to grow these goods in any substantial quantity. It imports shoes and shirts because they use quite a bit of labor in production and labor is expensive in the USA - shirts and shoes are too expensive to produce at home when someone else can make them much cheaper.

But there is the issue: because labor is expensive in the USA, and cheaper elsewhere, trade tends to reduce the size of the labor-using industries, and the number of these jobs in those industries. This is what some politicians refer to as the "exporting of jobs." Of course, the flip side of this story is also true. When a country trades, it also exports, and the industries that make those exports grow as well - this leads to job growth in the exporting industries - the "importing of jobs" if you will. There is a very good chance that there is a net positive job gain for each country trading as a result of this. After all, the industries that the country was strong in are the exporting industries, and they grow larger as a result of trade.

Still, those who lose their jobs as a result of trade are typically not in a good position. There is a fairly clear argument for government intervention here. Job retraining, education credits and incentives to create new jobs in the areas that are most distressed would make sense. However, governments have a long history of not doing this well, and the lack of support for those damaged by trade generates strong anti-trade sentiment. It is possible that such sentiments gain enough force to unravel and reverse the movement toward freer trade in the world.

3.5 China, India and Brazil

3.5.1 China should be a successful economy

China has always been a large and important country. With over a billion people, China's political and military position has been strong, though their economy has lagged. This may soon change. We have come to see China as a major manufacturer of household goods over the past decade. They dominate toys, small household appliances and some categories of electronics. It seems likely that they will become dominant in textiles and possibly large appliances. Their economic advantage, at the moment, lies in their enormous labor abundance relative to the rest of the world. It is reasonable to expect that advantage to continue to hold for most of the century.

China's remarkable growth in exports is even more impressive when one considers how briefly the expansion in exports occurred. From 1983-93, China's exports grew linearly, from about $25 billion to about $100bn. The growth in exports appears nearly exponential for the next decade, roughly from $125bn in 1994 to over $400bn in 2003. While exponential growth in exports cannot be maintained for a long time, by the time this growth is finished, China will be a major trade partner with many nations.

Most of us think of China as largely a country with a large labor supply and not much else. Perhaps because of the communist structure we think of the country as backward and not able to catch up with our technology. This would be a mistake. You should recognize that China was among the most technologically advanced countries in the world in 1600. Both gunpowder (in Chinese use about CE 1200) and ceramics that produced a strong lightweight result (porcelain starts about CE 600), were for the most part their technology. They enjoy a culture that is much older than that of the West, with a rich tradition of innovation and invention. They are rapidly deploying new technology now, and there is every reason to believe that they will become important in modern manufactures by the middle of the century.

There is even more to China than exports. As China's output grows, so does its income. There is already a small emerging middle class, and provided China stays on its current growth path, the middle class will be getting much larger. This means that China will be a large and fast-growing market for goods and services. With a billion people, that will be a very, very large market. Many western companies are well aware of this, and are already building a foothold in China's economy. With a billion people it might seem that a good industry to be in would be food-related. Economic evidence suggests the contrary - food-related demand is likely to grow quite slowly. Certainly finished food products and food services (like restaurants) will enjoy substantial growth, but not food demand itself. The growth industries tend to be more what people desire than what they need. Automobiles, nicer clothes, better and larger televisions and stereos, etc. Moreover, the large export surplus has been maintained in part by a fixed exchange rate. At some point, the exchange rate will have to rise, making foreign goods imported into China very competitive. As long as growth rates stay high, there will be many new opportunities for risk-taking business in the coming century.

3.5.2 India has an uncertain future

Much of what can be said about China can also be said about India. While India is not structured by communism, it has social structures that are quite rigid and may act as an impediment to

3.5. CHINA, INDIA AND BRAZIL

rapid adjustment to economic problems and opportunities. Within the last decade of the 20th century, India began to open their economy to trade, and the resulting changes have been quite dramatic. Like China, they have a population in excess of one billion, and they enjoy a significant labor advantage over much of the rest of the world. At the moment, they appear to have an education and technology advantage over China, and their industrial output reflects that advantage.

India's trade growth has been fairly linear over the past decade. In 1993, they exported about $18.4bn, by 2003 their exports were valued at about $50bn. A steady growth to be sure, though not an exponential one like that of China. Unless there are dramatic changes in government policy, the linear trend would be expected to continue.

The discussion of "outsourcing" in the United States is often about the flow of some jobs to India. This is especially true in the software industry, where India has a concentration of skilled engineers. India also has well trained doctors, who provide diagnostic services to other countries. Perhaps the biggest surprise is lawyers: because of the shared colonial past with Great Britain, India has a similar legal structure to Britain and the United States, so it provides legal services to other countries as well. This is likely to continue to be an important issue for the rest of the century, but it will not be exclusively an issue related to India. India will have competition for these types of jobs from East Asia, and probably China as well.

India's prospects are not as certain as China's. Because the remaining social aspects of the caste structure in their society, and the relative instability of the countries nearby them, it is hard to know if India can maintain its economic growth consistently during this century. India has no wealthy trading partner in its region, and may end up isolated, should trade break into regional blocs. If India is to perform well during this century, they will have to deal with their social structure, their neighbor's problems and their commitment to trade.

3.5.3 Brazil has a geographic advantage

Since we a discussing large countries that appear to be up and coming, we should also point out the Brazil has similar possibilities. Their population is not quite as large, and their usable land area is smaller, but they have a positive outlook in their people, and have maintained a fairly open economy for the past 20 years. They have strength in the steel industry, and are becoming strong in automotive parts and regional jetliners. Their neighbors are either fairly stable or very small countries relative to Brazil, so Brazil tends to have greater control over its destiny.

Brazil also has better access to the US and other western markets, as it is located in the West, and has many ties to western countries. If the world trading structure breaks into blocs, there is a good chance that Brazil will fall within the North American bloc. That would give Brazil access to a sizable market for its exports and would provide North America a vibrant semi-developed market for its exports.

To extend the analysis, should the world trading structure break into blocs, China will likely fall into the East Asian trade zone, where there are several developed markets, and a sizable population for their exports. India is in the most difficult position, as there are no sizable and well-developed countries in its region. A world trading structure that is dominated by large blocs of countries trading with each other but not much outside the bloc would probably leave India with limited access to the large well-developed economies. This means that India has greater risks for its export markets over this century.

3.6 US financial issues and a risky US dollar

There are also financial issues that are related to trade and the worldwide political strength of the United States. After the fall of the Berlin Wall in late 1989, there was talk of a "peace dividend." Much of this referred to the decline in the need for military expenditures, thus freeing resources for other uses, and the possibility of reduced taxes. This dividend should have been felt worldwide

3.6. US FINANCIAL ISSUES AND A RISKY US DOLLAR 57

in the same manner as the US, so most countries should have felt reduced tension and been able to allow their military expenditures to fall. It is quite possible that the relatively strong growth in the world economy is connected to the "peace dividend."

There is, however, more to the collapse of communism than a "peace dividend." After the only major military power to challenge the USA fell, the USA became the paramount leader of the world. The implications of such a position for a country are wider than the scope of this book. We can focus a few of these implications, starting with seigniorage. When you hold a US dollar in your hands, you are holding a part of the debt of the Federal Reserve, which effectively translates to debt of the federal government. This is in large part what makes the US dollar such good money. It has no intrinsic value of its own, only what you agree it is worth. This debt you hold, however, has a special property: it has no interest rate. The longer you hold the money, the longer the interest free loan you give to the US government. Other countries hold US dollars too. Their central banks hold the currency to ease international transactions (such as trade) and to pay interest or debts to foreigners. The more US funds they hold, the more interest free debt the US government enjoys.

Over the prior decade, the US had experienced high interest rates and a falling level of inflation, which made the US currency quite stable. So the dollar was already a desirable currency to hold by central banks. Now we need a concept called "seigniorage." In its classical definition, this term refers to the difference between the cost of producing currency and its face value. So, if it costs the mint $.05 to produce a US quarter (worth $.25), then if you hold this quarter (say you sock it away because it has your state on it), the government gets $.20 of seigniorage. More generally, currency that is held, but not used for purchases (or not circulated) can be regarded as interest free debt. Once the USA became the strongest military and economic world power, the desire for holding US dollars by central banks intensified. As holdings of US dollars increased by foreigners, the level of seigniorage increased, and more US debt could be financed interest free. From the US point of view, that's quite a "peace dividend"!

At some point, perhaps by the middle of the 21st century, US power will erode. It is quite possible that the formation of European Monetary Union and the Iraq war have already eroded some of that power. As the political power declines, the currency's suitability as a reserve currency declines as well. Once the flow of dollars reverses, and the US begins to see a flow of US dollars from abroad, the result will very much depend on the level of US deficits and the FED (Federal Reserve) response.

High US deficits make the FED's job much more difficult. The FED has to decide how much of the deficit to finance with bonds and how much with issuing more money. It is generally accepted that if money supply grows too quickly, the result is inflation, with roughly a 3 1/2 year lag. But if the Federal Reserve chooses to keep inflation in check, then they must bond-finance the deficit. Since they must compete for lending capital with all the other possible borrowers of such capital, there is a tendency for interest rates to rise. These other borrowers include mortgages (for housing), corporate bonds, student loans, automotive loans and credit card debts. The higher the deficit, the more that must be financed, so difficult decisions must be made about the choice between inflation or high interest rates.

We've been lucky over the past few decades, as foreigners have been willing to buy our bonds, thus allowing us to finance our deficits with foreign capital. This is not likely to last for much longer, though it might hold up for the next few decades.

Once the easy access to foreign credit ends, the US will face some difficult choices. It must either balance the fiscal deficit, allow interest rates to rise or allow inflation to rise. The political will to either raise taxes or cut spending appears to be limited, and because of the built-in inertia in government spending, the resulting fiscal deficits are likely to be around for a long time. This leaves us with an interesting scenario: the lack of foreign capital will force US interest rates up to cover deficits and perhaps some debt payments. The politics of this will make this uncomfortable situation only last for a short time. So the US will likely inflate its way out of this.

A new round of inflation sounds bad, but there is something

unique about the US debt structure. Nearly all US debt is denominated in US dollars. The federal government will never default on such debt, but it has incentives to reduce the debt's value. A sustained inflation of 10% per year can wipe out more than 4/5 of the principle on the debt within a few years. An inflationary process also tends to drive the value of US dollar down against foreign currencies, which makes US goods and services more desirable on foreign markets. This tends to help the US job market, at the expense of foreign job markets. From a political perspective, inflation seems to resolve two problems at once. That's a fairly strong incentive for the federal government to manage this issue with inflation.

There are clear choices you can make to protect yourself in such a scenario. There are two types of domestic investments that should help: inflation-indexed investments and those investments that have real underlying value. Real estate is a typical choice, though there are those who claim precious metals are very good for holding their value. The chief problem with precious metals is that they are subject to wide price fluctuations, as they are driven by supply and demand. If you are unlucky enough to buy near the peak price, you can lose your principle, and fairly quickly.

The other major choice for investment in an inflationary scenario is foreign assets. Here you are taking advantage of both foreign rates of return and the extra boost provided by the falling US dollar. As the US dollar falls, the foreign rate of return in US dollars is amplified by the depreciation of the US dollar. You should be cautious about where and what you invest in. While some countries may have very high rates of return, they are often risky places to invest, and may well have inflation problems of their own. If you invest in primarily well-developed countries, your portfolio should be fairly safe and can hold its value in US dollars.

3.7 World financial flows are important too

While US financial issues matter more to US citizens, the flow of finance in the world still matters. Capital and money have flowed in many directions over modern world history. These flows are strongly connected to trade activity. Part of this story is a result of each country's trade deficit or surplus, and part is due to the accumulation of assets that result from the international investment choices made by the citizens of each country.

The rather large trade deficits that the United States has had over the first 5 years of this century have meant that some manufacturing jobs and industries have been lost to the rest of the world. But that is not the whole story. These large trade deficits have also brought a large amount of foreign investment into the USA. The inflow of foreign funds has kept the interest rates down, which in turn has made it easier to borrow for both consumers and businesses. Consumer borrowing has helped maintain the economy-wide demand for goods and services. Business borrowing has kept up investment levels that have also allowed employment to remain high. It is clear that the US has gained overall from this, though these gains are not distributed evenly. Jobs with low skills have typically been lost, but those with high skills and education have been gained. So while there has been a shift in the type of jobs the US economy offers, the number of jobs has remained fairly high.

The shift in job and business types that flourish is important, and has been repeated throughout the world in most countries, particularly in countries that are engaged in trade. If we were to stop there, we would have a sense of how industries (and the jobs that go with them) grow or die off over time. Most trading underdeveloped countries have or will develop manufacturing industries that use labor as a major input. As a result, manufacturing jobs will grow in these countries. China is the current leading example of this trend, though over the century this will probably shift around from country to country. The reason for this is that countries that produce large trade surpluses also produce large amounts of foreign funds that can be used to re-invest in the developing economic structure of the country. As these re-investments take place, the type of industries change, typically becoming more dependent on capital and less dependent on labor. China has a long way to go on this road, perhaps 25-50 years before a shift toward capital intensive production overtakes labor intensive production. By then, other underdeveloped countries will have become the big producers of labor-using manufactured goods.

3.7. WORLD FINANCIAL FLOWS ARE IMPORTANT TOO

And yet, there is still another dimension to where the world is headed. It is the purely financial dimension. China's large trade surpluses with accumulated financial funds to go with them are likely to be destined for re-investment in China. This is likely to be true for most rapidly developing economies. But what about Saudi Arabia, Kuwait and other Middle Eastern economies? Their large trade surpluses with the USA and much of the rest of the world are not based on manufacturing. Indeed, their productive effort is quite low for their given output and large surplus. These countries have enjoyed their surplus and wealth because they happen to be living on valuable land - land that contains oil that is easy to extract. What have they done with their surplus?

To be sure, some of the surplus (and financial funds) that Middle Eastern countries have enjoyed has gone back as investment in their own country. Some has also gone back as consumption. But a large portion has become investment in other countries. These investments include US national debt, both short and long term. Some of these investments have gone into purchases of productive resources of other countries. The ownership of the world's productive resources is changing, and the shift this time is ownership by those in oil-rich countries. This is not likely to abate any time in the near future. Even if a new low-cost energy is developed in the near term future, that technology is not likely to be available to underdeveloped countries. So they will become users of oil, and oil rich countries will continue to enjoy large trade surpluses, which will provide them financial funds with which to buy other countries' productive resources.

The accumulation of world assets by oil-rich countries will probably accelerate over the century. When crude oil prices rise (even if the rise is temporary), these countries enjoy larger trade surpluses and more financial funds to go with these surpluses. As known oil reserves decline, oil prices will likely rise, which will bring more funds to oil rich countries for the same quantity of oil (or the same flow of funds for a lesser quantity). There is no obvious point at which this will end during the century, though the development of oil alternatives and newly discovered reserves will dilute the current oil-rich countries' share of world assets. This means that at least some part of the financial flow during the 21st century will be rooted in the oil-rich countries. At the moment, these are countries in the Middle East, Venezuela, Russia, Algeria and Nigeria. If we include the non-traditional reserves of heavy oil and tar sands, Canada moves into the list, and Venezuela becomes a very large supplier of raw petroleum product. However, these non-traditional reserves can't provide the same financial gain as light sweet crude because they come with considerable refinement cost.

The changes in ownership of assets are not necessarily something that will cause damage to the world economy. As noted before, these changes have occurred in the past, and today there is a rather large number of foreign-owned assets in virtually every country in the world. Many foreign-owned assets cannot be moved (for example, a building or other real estate), so who owns them may not be very important. Who owns moveable assets (like money and capital equipment) may be quite important. As more of an economy's assets are foreign owned, less strategic decision-making occurs inside the economy and more from the asset owner's country. Still, the aim of the decisions are the same - to make a profit, so who owns what assets has limited effect on how an economy operates, although foreign owners can and sometimes do make decisions on a political basis. What does change is where the profits go. Foreign capital owners will want to reap the benefits of their investment, and they do this by repatriating their profits home. That places downward pressure on the host country's exchange rate markets (or their balance of payments if they have a fixed exchange rate).

In the middle part of the 20th century the United States was (on net) an owner of foreign assets. By the late 1980's this shifted, so that the USA was (on net) a borrower of foreign assets. This effectively accelerated over the next 15 years, so that as of 2005, the USA is a major borrower of foreign assets. This cannot continue indefinitely, so at some point this will reverse. This reversal is connected to the discussion of the prior section, and may amplify the effects of a decline in foreign financing of US trade and fiscal deficits. It is hard to know precisely how this will effect the economy. Technically, all that is required to finance the outflow of foreign assets is a trade surplus. The link from trade balance to fiscal balance is not absolute. In practice, however, they often approximately mirror each other. So now we have even more support for the suggestion that the so-called 'twin deficits" of the US (fiscal and trade deficits) will come to an end fairly early in the 21st century. Given historical trends, it will probably take three decades of "twin surpluses" to rebalance. If the USA can run a fiscal and trade deficit until 2010, then it will run a surplus in both areas until about 2040.

One possible means that the USA could gain from this would be to become the developer of alternative energies. This would be a strong export market for the entire 21st century, and could be a major source of exchange market revenue. Relying on declining oil imports (due to dwindling supplies) probably won't work, as the cost of those supplies will be rising. However, energy is not the only area the US could lead. It already has leads in technology, entertainment, pharmaceuticals and financial services. There appears to be no shortage of talent, ingenuity and willingness to take risks in the USA, and there is every reason to believe that these attributes will remain an important part of the US economy throughout the century.

3.8 What to expect in US economic growth

If we consider both history and the Neoclassical Growth Model, there is reason to believe that the United States will enjoy a roughly 3.7% growth rate annually, on average over the next century. By no means should you expect this to be smooth. History suggests considerable variations in the growth rate of the US, some quite unpleasant. The most well-known downturn occurred in the 1930's, a period commonly referred to as "The Great Depression." It was followed by a very high growth rate that was coincident with the Second World War.

While those are certainly the most dramatic variations in the growth rate, you can see from the graph that there is quite a bit of variation, and it tends to be fairly cyclical. The up and down movement of the growth rate is called the business cycle. The US business cycle is about 5-7 years in length, with a few outliers. There is no reason to believe that this will change in the future. You should expect to deal with recessions and expansions, and you should be concerned about a recession after the 5th year of expansion. The bigger worry is the possibility of a depression.

Could another depression happen? First, we should recognize that this is a real possibility in the future, and there are few indicators of when it might happen. Most economists believe that the Great Depression was precipitated by an unusually large financial crisis. While the popular view connects the Great Depression with the stock market collapse of 1929, most economists don't see it that way. There were 4 waves of banking failures from 1930-32. The banking failures were on a massive scale, and are probably the most important difference between the Great Depression and the more typical recessions that occurred before and after.

The Federal Reserve (FED) knows that a catastrophic failure of the US banking system may well lead to another depression. As a result, it seems unlikely that they will allow this to happen again. This argument is one of the key arguments behind the bailout of the savings and loan industry in the 1980s. So as long as it is possible to prevent a collapse of the US banking system, it will be a top policy priority to save it.

Just the same, it is possible that some unforeseen events will come in succession at a time that finds the US banking system weak to begin with. It has been suggested above that at some point US dollars held as reserves in foreign banks may be coming back at a rapid rate. That would place intense pressure on the FED to drain the system of cash, usually by raising the interest rate. Banks would be weakened by the lack of credit, and possibly by some poor investments in interest rate sensitive areas. Perhaps an earthquake along the West Coast or a major hurricane in the South would make business conditions even more difficult. A natural business downturn could gain momentum at this point, and precipitate a banking crisis. Couple this with a sharp decline in international trade (perhaps the result of a sudden rise in tariffs), and you would have a serious recession if not worse. It is hard to know what the FED might do, or even if it could make the right

3.9. SOME RESOURCES FOR INTERNATIONAL TRADE

decisions at the right time.

Before you run off to your local bank and withdraw all of your savings and stuff them in your mattress, you should know that the US banking system is easily among the safest and most stable in the world. All deposits are guaranteed and backed by the federal government to $100,000. The scenario above is purely speculative, and is only to suggest a way that such a crisis could occur. Simply put, for a banking crisis to occur, a number of nasty events need to occur with very unpleasant timing. These things are impossible to forecast. The fact that they do happen should be moderated with the fact that they are rare as well. At this point in time, there is no such event looming in the near future.

You should also know that asset markets (stocks, bonds and real estate) have had bubbles. Stock markets in the US have had two serious bubbles in the last century: the late 1920's and the late 1990's to 2001. Very large losses occurred for most investors, and very rapidly. There have been several real estate bubbles as well. As of 2005, it is believed that there is a real estate bubble in several regions of the US, particularly along the coasts. If you are investing for the long term, the bubbles are not terribly important. But if you have short-term investment needs they are very important. The fact is that over the last century, stocks have returned about 12.5%, bonds about 8%, and money markets (essentially US treasuries) about 5% annually. If you think the market is in a bubble, treasuries make sense to protect your principle. If you want to invest over a long time horizon, stocks make more sense, though you should use an indexed mutual fund to spread your risk.

3.9 Some resources for international trade

The definitive site for information about world trade is the World Trade Organization's homepage (http://www.wto.org). They do, however, have a point of view, so you should not rely entirely on their papers and presentation. Virtually every country in the world has a government site associated with international trade. When you want to know about a specific country's trade interests, it is best to go to their site and look around. I did that for China, India and Brazil. For example, I used the Ministry of Trade in India for information (and the numbers I reported).

I also needed information about foreign holdings of US bonds. Here you need to look at the central bank of each country, and again virtually all countries have a site for their central bank. They will often list their chief foreign currency holdings (most hold US dollars in large quantities) and their holdings of foreign bonds. The place to find all of this information is the International Monetary Fund (IMF). It is at http://www.imf.org/ and can provide you with both research and data on international financial issues.

Any discussion about trade, free trade or globalization is rife with politics. If you want a top economist's view of the topic, you should take a look at *In Defense of Globalization* by Jagdish Bhagwati from the Oxford University Press in 2004.

Both the United Nations Development Programme (UNDP) and The World Bank publish annual reports with many of the themes I've discussed here. The United Nations annual report is called the *Human Development Report*, and The World Bank annual report is called the *World Development Report*. Both provide extensive analysis and data. Since the focus of the reports varies from year to year, you'll want to look at many of them to find topics of interest to you.

If you want to look at a good textbook on International Economics, you have a wide array of good texts written for college students. My personal favorite is *The World Economy, Trade and Finance* by Beth V. and Robert M. Yarbrough, currently in the 7th edition (2006) published by Thomson Southwestern. I've used the book for my classes on and off for many years, and I find it clear and accurate for most topics.

Near the end of this chapter I also introduced you to US economic growth. A very good resource for information about the US economy can be found in the annual Economic Report to the President. A long term data series on the US business cycle can always be found there. The report is generated by the Council of Economic Advisors, who also do quite a bit more than US economic analysis. For my data on China's export growth, I used a report by N. Gregory Mankiw, *China's Trade and U. S. Manufacturing Jobs*, which

3.9. SOME RESOURCES FOR INTERNATIONAL TRADE

was given to the US Congress as testimony on October 30, 2003 as part of the Council's function.

CHAPTER 4

The Demographic Future

Now that you have a sense of energy and international changes and how they impact the economy, you need one more area that is critical for the future: demographic change. This is a bit like saving the best for last because this is the one area of forecasts that is fairly well known and stable. To put it bluntly, forecasting energy prices, international trade and finance are at best imprecise - demographics are much better, and are known with near certainty.

You are probably rolling your eyes by now, thinking this has got to be the most boring section of the book, and you are very tempted to skip the entire chapter to get to the fun part of the future. Don't do it. There is very little of this material that you know, and it is one of, if not the most important determinants of how the future will unfold.

Some of the demographic changes that will be coming you've heard about: mostly an aging population in the United States. This same change is going to be prevalent, perhaps more pronounced, in Europe as well. But population aging will occur all over the world, and will produce many results that will change the structure of production and asset allocation worldwide. These changes are discussed, as well as how societies will have to deal with changes in the structure of those working and how they support those that do not (this is called the dependency ratio). Because Africa emerges as the continent of opportunity by mid-century, we will cover one its greatest threats: HIV/AIDS. There is also one of the strangest twists yet to play out as a result of China's one child policy, that of the "missing girls" in the Chinese demographic structure. The idea that there may be limits as to how many people a country, or a world, can support is also discussed.

Even though you have a sense of an aging society, you have probably not thought through how this change will affect what goods and services are likely to be influenced. I try to provide you some ideas as to what might happen to the demand structure in the future. However, financial markets are also affected by an aging society. Both savings and investment choices may change in ways that reflect the need to protect assets and produce income, especially for those who are no longer working.

Demographic change will become one of the driving variables in the last chapter. You'll need this chapter to see why one continent gets a major opportunity while others are forced to adjust to a population that is either not growing, or even declining in size. Demographic change also figures prominently in how assets and even spending are influenced by changes in trade and energy prices. This chapter also helps you understand how civil wars come about, and perhaps how to help countries avoid them in the future.

4.1 An aging population

The population of both the USA and the world will age over the next century. Of course we are all familiar with the Baby Boom population segment of the US society. That creates issues that will be important in the next 30 years. The aging of the population referred to here is a wider phenomenon, and has consequences that are much broader. Barring an unforeseen pandemic, or a world war that wipes out a large portion of the world population, it is almost certain that the average age of the world population will rise. While world population has been rising over the past several hundred years, and particularly fast over the past 100, there is evidence that the rate of population growth is slowing down. This has

4.1. AN AGING POPULATION

probably as much to do with a natural maximum as it has to do with efforts by the United Nations, World Bank and WHO. Given a decline in the birth rate, and the fact that people are now living longer, it is a fairly clear result that the population ages. This does not mean that we should expect a large portion of elderly worldwide, but rather some groups of countries experience a graying population while others enjoy an increase in the most productive age group.

An aging population is about the mean age moving up. While in the well-developed countries (USA, Canada, European countries and Japan) this means a sharp increase in the elderly population, the effect in the underdeveloped world (Latin America, Africa, and the Middle East) is quite different. For the underdeveloped world, at least for the first half of the 21st century, the size of the most productive age group, the teens to the forties, should rise proportionally to other age groups. This is part of the reason we should expect higher output worldwide, though higher productivity per labor hour should be the more obvious reason for the rise in output. A decline in the birth rate, especially in underdeveloped countries, would be a welcome relief. There is a fair amount of economic hardship that is a result of high birth rates and a young population. Should a decline in the birthrate really occur, many underdeveloped countries may have their first chance at economic stability in a century or more. Economic stability often brings political stability with it, so improvement in the political arena is possible as well.

In 2005, the United States is a fairly typical well-developed country with a flattening population-age distribution. The lower age groups, from 50-54 and below are about even, while the size of the age groups declines progressively as age increases, though this becomes larger for the oldest group at 80+. This is mostly due to the oldest group not being broken down into 5-year subgroups. This might be most easily seen with an age-sex pyramid graph, where the age for each sex is graphed on the vertical axis and the size of each group is on the horizontal axis. Take a look below:

However, by 2030, the United States has a relatively flat population-age distribution. Notice the even sides all the way up. This tells us

United States – Age-Sex Pyramid 2005

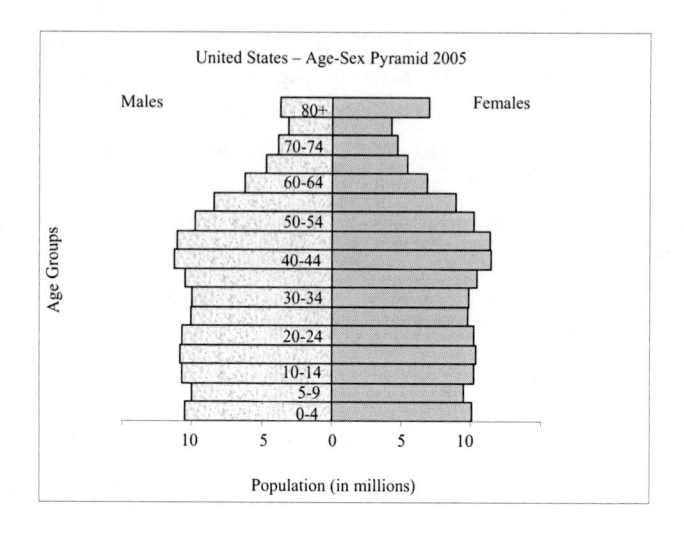

Figure 4.1: source US Census Bureau, International Data Base

that the population for each age group is about the same. This is the chief reason we are worried about Social Security now. When there is a large population of younger people supporting a small population above, the amount of funds being paid-in finances the funds being paid out. When a society becomes older, the structure of these payouts can no longer work. Take a look at the nearly rectangular population structure, below:

Given the shape of the USA in 2030, by now you are wondering why these graphs are called age-sex pyramids. Maybe they should be called age-sex rectangles. If all countries looked like the USA, perhaps they would, but in fact young, fast growing countries look very much like a pyramid. In fast growing countries the youngest groups in a society are the largest, while the oldest are quite small. Poor nutrition, abject poverty and bad living conditions mean few reach older ages. The people's response to this is a high birth rate to insure that some survive to childbearing ages. As a result, there are many children born per family, and this tends to expand the

4.1. AN AGING POPULATION

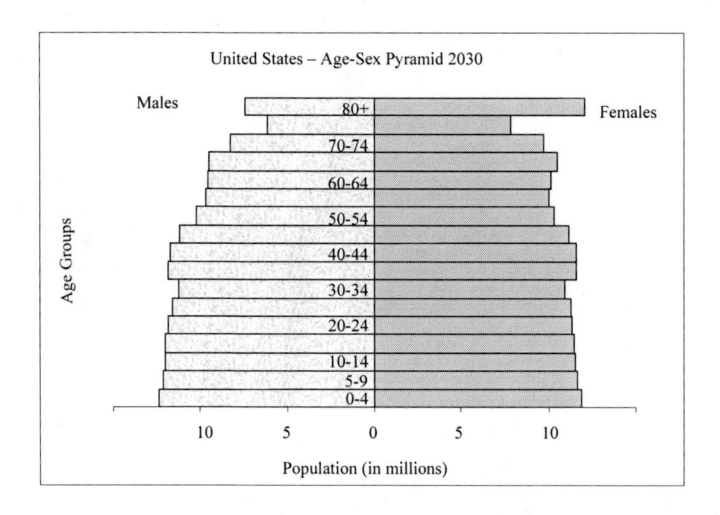

[Figure: United States – Age-Sex Pyramid 2030, showing Males and Females across age groups 0-4 through 80+, with Population (in millions) on the x-axis]

Figure 4.2: source US Census Bureau, International Data Base

population rapidly. Sudan provides such an example in Figure 4.3.

Here we see a shape that is very similar for most other underdeveloped countries in Africa. The distribution is very bottom heavy, suggesting a young tilted population. This sort of distribution tends to yield a fast growing population. With a large number of young people there will a large number of young, fertile women within a few years. This tends to create more young people and the cycle continues. We now know that the best way to break this cycle is to educate women. Educated women have income-producing alternatives to having children, and tend to delay marriage and childbirth. Education slows down the birth rate, and thus the population growth rate. The process of slowing birth rates has a flattening effect on the age-sex distribution of a country, a process known as demographic transition, which will be covered a bit later. Because it is difficult to keep a very young society supported with food and other necessities, this transition may well be welcomed. The issue for very young societies (or very old ones) is referred to

74   CHAPTER 4. THE DEMOGRAPHIC FUTURE

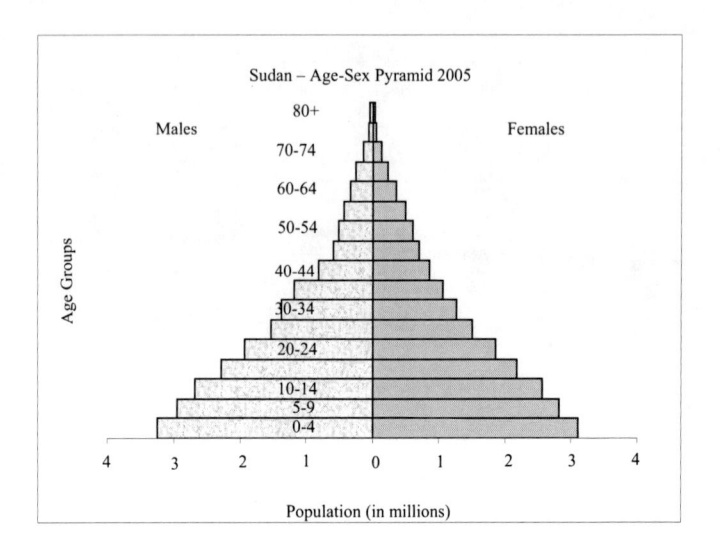

[Figure: Sudan – Age-Sex Pyramid 2005, showing population in millions by age group for males and females]

Figure 4.3: source US Census Bureau, International Data Base

as the dependency ratio.

4.2 A rising dependency ratio

Societies contain a wide dispersion of age groups, with the majority of the groups concentrated in the working ages (age groups 20-65). The working groups are typically responsible for producing the resources that the entire society uses. This means that the working groups must create a surplus of output that is then utilized by both the youngest and the oldest members of the entire society. The measurement of this effect is called the dependency ratio, and is defined by the fraction of those working (age groups 20-65) to those not working (both the very young and the very old).

There are three aspects of this that warrant further attention. The first is that the United States is not the only country that will experience significant aging of its population. Virtually all of western Europe will have this as well. The last decade in Japan

4.2. A RISING DEPENDENCY RATIO

(1990-2000) in many ways is a hint of the older orientation in age structure to come, and they did not weather it very well. Japan experienced a protracted recession and very weak growth for much of the decade, though the age structure was only part of the problems Japan faced. However, an increase in the oldest segments of the age structure appears to make most economic policy less effective, as retirees spend and save on a different basis than those working do. This is known as a life-cycle effect, and while it is well researched, new policy approaches will have to be developed to manage this new reality.

So what's the big deal? Don't we expect people to get older? It's not that people get older, it's that the age structure of the population changes. The population age structure tends to flatten out; the number of people in each of the age groups is about the same (like the USA in 2030). When a country's age structure is bottom heavy (more young than old) it is relatively easy for the younger working group to support the older non-working group. But in a flatter age distribution, this becomes more difficult. The older group, because it is larger, requires more resources to maintain its living standard. At the same time, the younger working group is smaller in relative terms than it was before, so it is more difficult for this group to create resources needed by all groups.

The dependency ratio in virtually all well-developed countries will be rising over most of the century. This goes far beyond the Social Security issues you've heard about. Certainly, government inter-generational transfer programs like Social Security will be in trouble. But larger issues will also come into play. Because those who are retired will be drawing down their savings, and they are a larger segment of society, the national savings rate will fall. And more resources will go into goods and services that address the needs of those retired. Health care is a big worry, but so are housing, transportation, recreation and other aspects of the needs of the retired. It seems unlikely that those working will suddenly increase their savings rate, or that economic growth will be so vibrant that the savings rate is lifted. Instead, some new structure will have to replace the current structure to make this all work. No one knows what that structure will have to be. However, it will probably include an older retirement age, and some form of reduced Social Security or its equivalent. Let me remind you again, this is not just a problem for the United States, but for Europe, Japan, Canada and Australia as well.

While most of what we hear about in the newspapers is about soaring health care costs and looming Social Security insolvency, there is more. Remember, in most if not all western countries the number of retired workers will rise, and rise proportionally in their societies. There are several facts that warrant attention. First, this group tends to be politically active - and by this I mean they tend to vote. So they will be a major political force in each of their countries. Second, they are retired, not necessarily frail or elderly. This means a changing market, and changing marketing tactics. These people need a different set of goods and services, and because they represent a large group within the society, they will be served, possibly being the target market group. They are older, living on pensions and accumulated wealth, and may well be the richest segment of society. Those businesses and industries that address their needs are the growth industries. And lastly, yes, because they are older they will need a higher level of health care. It is not unreasonable to suggest that health care expenditures will rise sharply in these countries. It is difficult to know how these economies will cope with this problem, as western Europe has a public health care system in place and this will strain their systems. The USA has a limited public health care system (Medicare and Medicaid), neither of which is up to the task of dealing with this issue. Social Security is also at risk because there will be fewer folks paying in, and more drawing from the system. This was not imagined when the system was started, but it is recognized as a problem now. The sooner a solution to the Social Security and health care issues is found, the smoother and easier the transition to an older society will be.

Moreover, the dependency ratio issue as it exists in well-developed countries will affect underdeveloped countries within 50 years. This is a bit of an odd flip - right now the dependency issue for these countries is concentrated in the youngest age groups; there is a large number of youth relative to working age adults. The ratio

4.2. A RISING DEPENDENCY RATIO

will fall over the next 30 years as these young people get older and become productive adults. Then, as they age, the demographic age structure will flatten out, and the dependency ratio will rise again, but this time the effect will be from the aging population. The age issues faced by well-developed countries now will be a problem for undeveloped countries by mid-century. But they will have to deal with the problem using considerably smaller resources. For these people, retirement will not be an option.

As already noted, the United States is not the only country that will experience these problems. In fact, most of Europe will enter the aging population skew earlier. In some cases, this will actually result in an inverted age-sex pyramid. By 2045, Italy will have a structure that is top-heavy; that is more folks will be older than younger. Figure 4.4 is their age-sex pyramid for 2045:

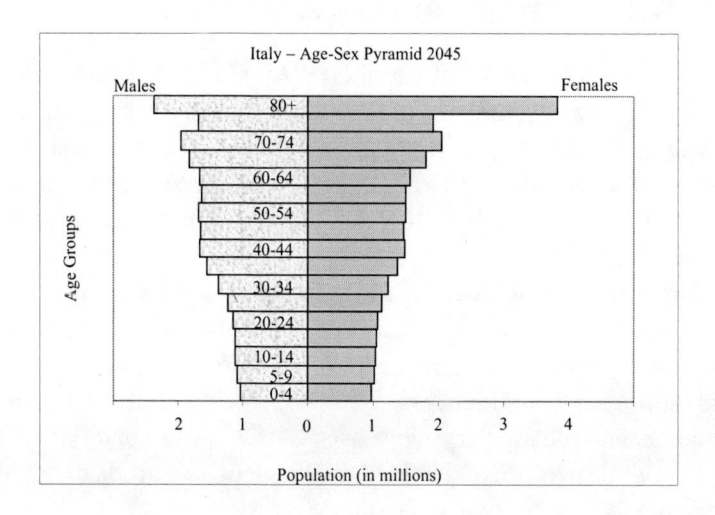

Figure 4.4: source US Census Bureau, International Data Base

Again, assuming that there are no catastrophic events, like a world war or a pandemic, the world as a whole will persist in this relatively flat age structure for the rest of the century (2050-2100). You have already seen that there is some evidence from European countries now that the structure might even become a bit top heavy. If medical science can extend our lives further, so that life expectancy rises to say 100, the age structure almost surely will become top-heavy. Should this become the case, retirement at age 65 will be too early. However, presumably because people will be healthier at that age, they may not feel much like retiring anyway. This would suggest that older workers might stay in their jobs longer. Given the improved health conditions, people should remain productive well into their 70's, maybe even their 80's. This does not mean working as a greeter at the local megastore. It means remaining in their regular jobs, drawing their full salary or wage for many years after age 65. A longer life expectancy combined with an economic need to remain working may even provide for two or more careers in a lifetime.

4.3 The damage of HIV/AIDS

The second aspect is the effect of HIV/AIDS. While this virus kills many in nearly all countries of the world, its greatest damage is occurring in Africa. Approximately 75% of the AIDS-related deaths occur in Sub-Saharan Africa. Even worse, these deaths occur in the population segment 20-40 years old, and typically among those with the higher education and skills in those countries. For an undeveloped country to sustain these losses is a damage of staggering proportions. It is not hard to see that many of these countries in the early development process are going to have a difficult time maintaining their momentum. They are likely to find development of their economies a very slow process for quite some time, and may not stabilize until the HIV/AIDS virus is cured, or at least controlled.

If you think this is their problem, you are only right to a certain degree. A weak Africa can't generate sufficient income to take care of themselves, and requires resources of richer countries to assist them. Prolonged economic weakness can also lead to civil war, and the corresponding terrorism that comes with it. By contrast, a strong Africa could expand production at a rate that generates a national surplus. This would provide the world with a source of labor-intensive exports and later an export market. A strong,

4.4 China and the missing girls

The last aspect is that of the population structure of China. This is one of the most fascinating issues about to surface in the near future. Because of China's one child policy, and the unfortunate preference for male children, one of the most unusual scenarios is about to occur in China. The gender ratio of males to females will be about 60-40, favoring the males. This may be the first time in history that such an event has occurred. There have been many times in the past that the ratio has strongly favored females. The most notable was post WWI Europe, when large segments of the younger population of European males were lost to the war. But this is different.

This problem is known as the "missing girls" issue, and is associated with the one child policy of China. This can be seen by looking at the number of males per hundred females in the lowest age group: in China (for 2000) it is 114.1 males per 100 females, by contrast, in the USA (2000) it is 104.7 males per 100 females. It appears to be mostly the result of sex-selective abortions, though more unpleasant methods may have been used to reduce the number of female babies. The end result is a society with a very skewed gender distribution favoring the males. A strongly male ratio surely has economic and social consequences. The position and status of women is likely to improve, possibly sharply and substantially. Based on history, one would guess that the highest concentration of unmarried males would be in the poorest income groups. It is possible that China will effectively import women from other countries. Given the male's tendency for aggressive behavior, it may be that China will become aggressive overall, and a war with nearby countries may be a concern.

To help you see China's male skew in their population distribution, here is the age-sex pyramid for China in 2030:

If you follow the 40 million marks all the way up, you can see how this gender skew looks in China. The gender skew does not get

CHAPTER 4. THE DEMOGRAPHIC FUTURE

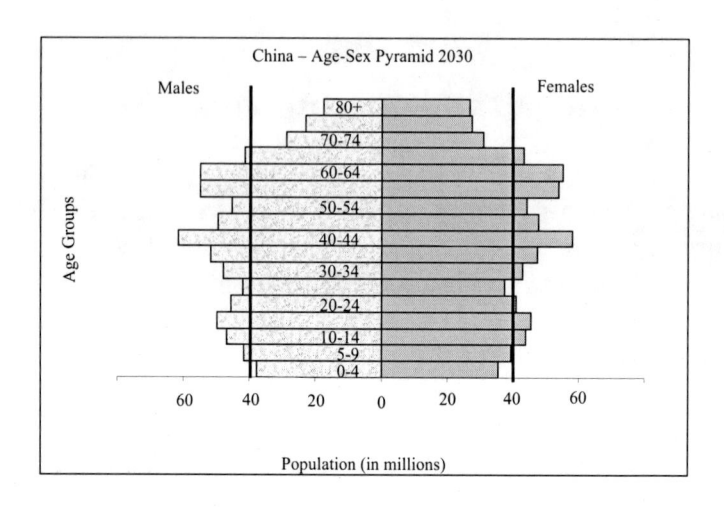

Figure 4.5: source US Census Bureau, International Data Base

back to normal until you reach the 70-74 age bracket and after. India also has a skewed ratio of females to males, as of 2005 at 106.5 males to 100 females. This ratio varies considerably throughout regions and ethnic groups, and for a few groups (those in the 25-35 years old age range) it does get close to China's male-dominated ratios. Based on the pattern for all age groups in India, it would also seem that this preference for males has been present for quite some time, so they have made this work in their society. Their population is also still growing fast, so the males can still draw from a large female population. While that suggests that China can make it work as well, the shift for China is very sharp, and the ratio is very skewed.

There is a significant marketing shift at work here as well. Because of the relatively large size of the male population, male-oriented products and services will become the hottest markets to be in. To be successful in China may require a revamping of product lines and marketing approaches. Men's grooming products and their known interest in work tools may find a huge market opportunity. Women's power position in the marketplace should drive their products decidedly upscale. The style of marketing will have to change as well, since the largest market target will be men. The imbalance may also generate a significant shift in cultural values, possibly giving girls an equal status to boys as babies. India's experience may give some hint as to what we should expect in China. The Chinese will need to get the ratio of girls to boys back to normal for future generations or there may be demographic damage later when the age structure flattens out.

4.5 Aging societies change what's in demand

The higher mean population age also changes the nature of goods and services that are in demand. The demand profile would tend to move toward a wider array of goods that are needed by people who maintain housing units. It should tend to move away from foodstuffs and other goods oriented to very young populations. Should a middle class emerge in underdeveloped countries, a vast new market would appear to absorb output from the housewares, automotive and financial services industries. Since most underdeveloped countries start the development process in light industrials, particularly textiles, we should expect these industries to gravitate toward these countries. Those countries currently dominant in light industrials will likely move up the development ladder.

The fact that world population will increase during the next 100 years would imply to most people that food prices will rise, and that the foodstuffs industry would be a good place to invest. However, well over a hundred years of economic research supports what is known as Engle's Law, which suggests that the demand for food does not rise as fast as income. The data goes further: the demand for food rises very slowly compared to income. So, even if the world's output rises sharply over the next century, one can reasonably expect food industry profits to remain fairly stagnant. Chances favor that the foodstuffs industry will remain very competitive in structure, and foodstuffs will remain largely a commodity-based industry.

Economists define a luxury good as one whose demand grows faster than income. In some sense this is Engle's Law in reverse. The idea is that certain goods tend to become more desirable as people enjoy more income. While not an economic law, it is well known that these goods do exist. Modern examples of these types of goods include large automobiles, large televisions and home theater equipment. Industries that make products or services that have this attribute are typically good investments because they have fast growing demand, so there is much space for profitability. It is not possible to know what these goods will be over the next century, though if you know what types of goods to look for, then you will know what industries to invest in.

4.6 Carrying capacity: resources at the limit

One of the most significant demographic issues that we will deal with in this century is "carrying capacity." Carrying capacity refers to how large a population can be sustained with given resources (mostly food and energy). With world population currently at six billion, and a projected nine billion by mid-century, how much the Earth can carry is a very important question. Unfortunately, there is no consensus on this issue. The estimates range from two billion to 36 billion people. That's hardly helpful.

Part of the problem is determining how well people live at various levels. If we assume people live at the level of those of us in North America, we arrive at the figure of two billion. Let me put that another way - If the population of the world was two billion, all of us could live at a high standard of living, like those of us in North America.

If we make the standard of living much lower, say near subsistence levels, like much of the developing world already lives in, the carrying capacity of the world is much higher, around nine to ten billion, maybe higher. None of this corrects for the natural drawing down of resources that are non-renewable (for instance, oil), though many estimates do include technological improvements.

It is, however, very much worth noting that world resources are not equally distributed, nor equally consumed. Western countries

4.6. CARRYING CAPACITY: RESOURCES AT THE LIMIT

(USA, Canada, Western Europe) have been lucky. They came to the world's table when there was little competition for resources and had an impressive amount of resources of their own. These days appear to be numbered. Within the first 5 years of this century, we have already felt the effects of a more productive China.

Issues related to carrying capacity for the world also are relevant for individual continents and countries. Countries that are rich in natural resources and particularly arable land tend to have a greater carrying capacity than those that do not. In terms of arable land, the United States and Argentina are among the richest in the world. The United States also enjoys the largest fresh water sources (the Great Lakes) in the world, which is approximately 1/4 of the world's supply. Europe as a whole also has much in these areas, but it is split up into smaller individual countries. Africa and Australia are the poorest continents, and given the large population of Africa, it is the weakest of all.

Africa's relatively weak resources and large population places it at risk. Many of the people on the continent already live at the margin, and even small environmental changes can have grave effects. A drought year quickly translates to starvation; too much rain to flooding and water-borne diseases. The combination of a large population and a relatively small carrying capacity means that Africa will likely need assistance for the entire century, and it is hard to know if the world is willing to commit to assisting them for that long.

South and Central America, while experiencing population pressures as well, are on much better footing. Their natural resources keep most of their population above marginal existence. However, distribution of income problems mean that many continue to live at subsistence, and this will likely be an issue for much of the century. Venezuela, with its oil reserves, is a likely winner in the first half of the century, and with its other natural resources, it is likely to be successful in the second half as well. Brazil, Argentina and Columbia are also well poised in terms of resources, but political and income distribution problems will need to be solved for those countries to become success stories in this century.

Asia also has population pressures, but it also has natural resources to deal with it. Some countries, notably Japan and Taiwan, are relatively weak in critical natural resources, but have successfully traded away this problem for quite some time. Others, like India, Pakistan and Bangladesh are near or past their carrying capacity and will have riskier futures.

Any country that has exceeded or nearly exceeded its carrying capacity is certainly at risk. But there is yet another element of risk that tells us about the future for these countries. The problem that shows up when population growth slows down is the resulting population bulge that appears. With fewer people that are older, and fewer that are younger, there is a resource squeeze for the bulge group. Demographers call this demographic transition, and it has already started in many countries.

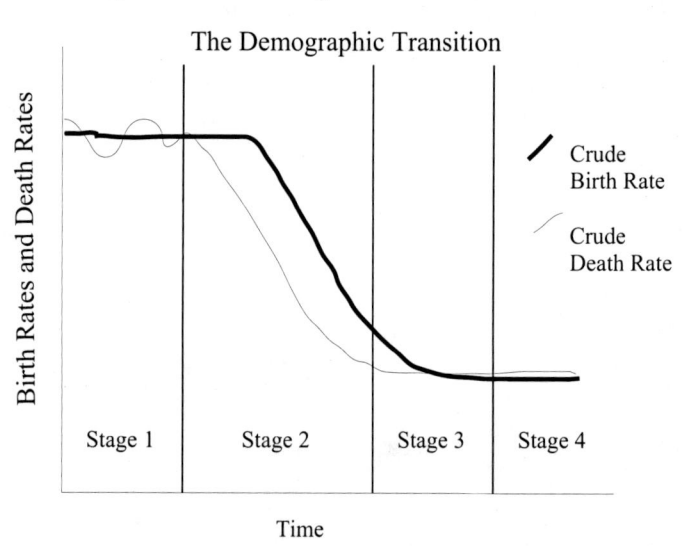

Figure 4.6: source: Stages 1-3, Sixth Sense, Oldham Sixth Form College

Transition generates considerable political, social and economic fallout. Much of the effect of transition can be traced to lack of jobs for the larger population group, and the corresponding lack of access to resources, goods and services. This transition tends to be

4.6. CARRYING CAPACITY: RESOURCES AT THE LIMIT

dangerous because the group where the problems are most concentrated is the teens to twenties. These groups are known to be the most likely to be emotional and violent. Several countries in Africa that have recently gone through transition are also those with violent civil wars. As more countries, particularly developing countries, go through transition, additional civil wars can be expected with their corresponding violence.

There is nothing good about a civil war. Countries lose sight of the production side of their economies, and tend to have sharp drops in per capita income and output. Besides the wartime horrors, the ethnic cleansing or genocide and the physical destruction of economic resources, there tends to be mass starvation and widespread disease. Because of the population size in underdeveloped countries, and the fact that many if not most live at the bare minimum, the effect of a civil war is likely to be of stunning proportions. The 21st century may well be marked by losses of human life that dwarf that of the 20th century.

There has been a suggestion that wars are good for an economy. First, this is, in general, simply not true. Economies that are in a deep depression certainly are helped by the government spending associated with a war, but therein is the key: any government spending would be helpful. Governments could just as easily spend on projects that are constructive, dams or highways for example. John Maynard Keynes told us this in 1936, and at least for an economy in a deep depression it is probably true. Civil wars are different with respect to where the fighting takes place. They decimate countries from the inside, and there is no victory that comes without a heavy price. No economy is better off as a result of a civil war.

When an underdeveloped country experiences demographic transition, a civil war is one possible outcome. But it need not be the only outcome. What seems to be missed in much of the literature on the topic is that the risky age group (late teens to early 20s) is also the group with one of the highest capacities for economic output. If the productivity of this group can be harnessed, what could be a terrible decline into a civil war could instead be the one of the most significant opportunities for economic growth. Proper planning and the support of markets are critical. Well developed countries have a stake in this - it might be argued that they don't have a stake in a civil war, but they clearly have a stake in tapping into a highly productive work force, and later a potentially large market to sell their products in.

The increase in the working age group as a result of the transition is sometimes referred to as the "demographic dividend." Several authors have pointed at both Southeast Asia and Ireland as successful at utilizing the demographic transition to grow rapidly. These countries planned for the increase in working age adults by expanding education and pursuing an exporting strategy that increased employment throughout their economies. Such a strategy may work for Africa as well.

You should note that both Europe and North America are known to have an increasing number of older, non-working age adults. This is why I've added Stage 4 to graph. Amoung other things, it is possible for a relatively old society to experience a population decline, as the birth rate is not high enough to provide replacement for those lost to old age. In order for this group of countries to maintain their supply of goods, they will either have to allow for increased immigration to boost their supply of labor, or they will have to increase their imports of goods that use labor in their production. In many ways this has already happened, as imports of labor-intensive goods flowed to both Europe and North America over the past 40 years, mostly from Southeast Asia. As Southeast Asia's societies grow into an older demographic phase, Africa's societies have an opportunity to become large exporters of these goods. This century will afford Africa a real chance at economic success. It is unclear whether they are ready for it.

4.7 Demographic change influences markets

Demographic change is one of the easiest things to forecast. Populations tend to be stable, with known characteristics (like birth rate, death rate, and age structure). Because of the stability of the distribution of the population, and what we know about the profile of demand from various age groups, we should be able to forecast, with a fair amount of accuracy, the structure of demand for a variety of goods and services well into the future. For example, we know that the United States population is aging, and will be getting older on average for the next 40 or so years. The goods and services that meet the needs of an older population should experience increasing demand. One of those services is nursing home care. We should also expect health care expenditures to rise. Retirement housing should be expected to rise, and this means that more than likely real estate prices will increase in the sunbelt states in the southern US, at least for the next 30 or so years.

There is more to this than the obvious demand changes. The types of movies and music that appeal to older folks will also be in greater demand. Clothing and foods will also experience structural changes. Retirement planners and financial services that support an older population should also see a rising demand. One must consider all of the effects that an older population has on demand structure in order to be in the right market position to gain from the shift.

What turns out to be even more interesting is that markets do not appear to anticipate these changes well. Perhaps markets are too focused on the short term to make these adjustments. It is possible that there is too much noise in the marketplace (too much information, much of it irrelevant) to see the bigger picture clearly. Whatever the reason, there is an indication that those who follow the demographic changes, and invest for the long term appropriately, should be able to do quite well in the market.

4.8 Places to look for demographics

Demographic information is well researched and easily found on the Internet and in books. There are some great sites for this topic. My personal favorite will give you an age-sex pyramid for nearly any country in the world, and put it into motion so you can see how the structure of the society changes over time. You can find this site at the US Census Bureau, IDB Population Pyramids, which is at: http://www.census.gov/ipc/www/idbpyr.html on the web. This is the Internet at its best, and doing something that books can't do -

providing a dynamic result. I built my age-sex pyramids from data collected from this site as well, though I imported the data into a spreadsheet and created the graphs using Open Office.

Of course, the US Census Bureau is one of the single best places to look for demographic data, but you might be surprised by how many other places there are to look.

I also used a book published by The Federal Reserve Bank of Kansas City in 2004, *Global Demographic Change: Economic Impacts and Policy Challenges*. For me, as an economist, I enjoyed the collection of papers, though I suppose to be fair, you might find it a bit dry. I drew from it when I discussed demographic transition and also the "missing girls" problem in China. It is a good place to start if you want an idea of the current thinking economists have about how demography will change things. The International Monetary Fund has a survey, *World Economic Outlook, The Global Demographic Transition* published in 2004. This publication does a good job of connecting demographic trends to trade and financial issues.

I used the SPC (Secretariat of the Pacific Community) Population / Demography Programme for some of my analytics in this section. The site is a remarkable location for information about the South Pacific Islands. You can find it at: http://www.spc.org.nc/ on the Internet. I also used http://sixthsense.osfc.ac.uk/geography/ which is at the Geography Department at Oldham Sixth Form College. In addition to a variety of topics in geography, you can find a very nice discussion of both the demographic transition and age-sex pyramids.

For discussion of the "missing girls" problem in China, you might take a look at China Population Information and Research Center (CPDRC) which can be found at http://www.cpirc.org.cn/en/eindex.htm in the Internet. This site also has a discussion of population aging in China, and many population policy papers.

If you want to do some reading on carrying capacity, you might try http://www.ecofuture.org/pop/info.html which is part of EcoFuture. There are many other sites as well, but you should be forewarned that most sites related to carrying capacity tend to have a

4.8. PLACES TO LOOK FOR DEMOGRAPHICS

point of view, and may need to be recognized as such.

If you are interested in Africa's status and development prospects, you should go to the African Development Bank (http://www.afdb.org). The site is the clearinghouse for information about African economic data and analysis. For the world economy as a whole, The World Bank should be given a good look. It is at http://www.worldbank.org/ and is one of the premier development research facilities in the world.

CHAPTER 5

Mix It Up: A Future With All In Play

So now we have the foundation of our story. There are three large areas that will be important determinants of the next 100 years. The movement of energy prices, the changes in international trade and the changes in demographics worldwide are the focal points. So much can happen outside of these areas that it seems that we might have nothing to say. That's not true. These three things are the driving forces for much of everything else.

Most countries go to war over economic issues. Sure there are other reasons for conflict, but without the economic reasons it's far more worthwhile to settle them without war. A combination of trade and demography are often at the root of these conflicts. We already know that energy (or oil in particular) is a reason to go to war. Wars are a real possible outcome of the conflicts that will develop over the century; but they don't have to be. When you know what's coming, and how to deal with it, countries can gain by working together.

There are a few things we ought to start with. First, demographic change is the strongest of the forecasts suggested by the prior three chapters. We can reliably forecast demographics at least 25 years into the future, and fairly well up to 50 years. The trade forecast is the weakest of the group, as we have no precedent for the current state of trade in the world. The world is currently trading as much as it ever has, and many feel this will continue, perhaps increase. Our forecast is different, mostly suggesting that the trade level has reached the maximum tolerable level, and will decline for most of the foreseeable future. The energy forecast is better than the trade forecast, but not by much. There are repeated warnings about the world running out of oil, as well as environmental damage from the use of fossil fuels. Based on the current data, there is not much evidence of the world running out of oil quickly, though environmental damage may well change our view of how we want to use this and other fossil fuel resources.

There is also a question of what affects what. As a practical matter, neither trade nor energy has much immediate effect on demographics. If either change income growth in a country or countries, there are long term effects, though they are probably fairly weak. So for the most part, this direction of causality will be ignored. The reverse is not true though, as demographic change has significant implications for both energy demand and trade activity. Energy prices most certainly affect trade, as trade affects energy prices. The timing of these changes also matters. So what happens, and in what sequence, is of great interest.

There are two approaches to the future here. The first takes the view that oil prices will rise over the course of the century, with most of the increase in the first 30 years. Because of this, energy prices are the chief dynamic of the century, and trade is damaged while the demographic changes fall into the background. The opportunity that Africa is expected to have gets missed, and while it eventually gets its chance, it happens much later. Western economies, (especially the United States) are seriously damaged as well. The world survives, but it is a much different place. Before you read this book, and perhaps even now, this is the future as you probably see it.

The second approach takes the view that oil prices remain fairly steady throughout the century, though it is punctuated by price spikes. These spikes are part of the dynamic of the century, but demographic change and trade are far more important. Africa gets it opportunity by the third decade of the century, before demographic transition ruins the continent. Still, there are great risks to the world financial system, caused by large fiscal and trade deficits, coupled with known demographic changes that effect the retirement systems of western economies. The financial disasters generated by these effects can be avoided, though it will require restraint now and in the future.

The truth is that you don't wholly believe that oil prices will remain fairly steady over the century, so it only seems fair to give you a future with rising oil prices first. Still, the analysis in this book suggests oil prices will remain about the same (on average) for the century, so this story is presented after section 5.1. In the end, you will decide if one story or the other is closer to what will happen. No doubt the future will play out somewhat differently than either story, but one of the stories should be closer to what really happens.

The next section offers you a future that you believe right now; one with rising crude oil prices and dwindling supplies. It is not a pretty future, but the price dynamics give results that are not as dire as you might expect. The last section of the book will provide an alternative future where rising oil prices are not dominant, because in part, they don't rise on average. Oil prices may spike occasionally, but the key to understanding the future lies in international trade and demographic change.

5.1 A first look - the future as you saw it

You've heard that we are running out of oil, and that the price of oil, and thus gasoline, is headed upward in some horrible unrelenting spiral forever. Since you probably believe that is what is going to happen, it makes sense to start you off with a future where oil prices rise. However, you should know that a rising price of oil is not the whole story, not even close. Changes in international trade, international financial flow and demographics are going to be at least, if not more important than changes in oil or energy prices. Consider the story unfolding in front of you...

If the price of crude oil was to rise consistently, and this was tied to some inevitable decline in the total reserves in the world, we would have a very interesting future. The "peak oil" adherents suggest that we have either passed our peak in the late 1990's or early 2000's. This suggests that for the next three to four decades, the dominant dynamic feature of the world economy will be the rising price of crude oil. What should we expect in such a world?

A steady increase in the price of crude oil would have a dramatic impact on the price of all energy. It should not take long for the price level of oil to reach a point where Canada's oil sands become viable. The problem is that to convert oil sands to synthetic crude, natural gas is needed as a heat source. So upward pressure on natural gas prices is a clear result. The price of heating oil will also rise, as it is a refined form of crude. Of course, gasoline, diesel and jet fuel will also rise in price. This will open market opportunities for near substitutes, like bio-diesel, ethanol and even wood and coal (as heating sources). Should prices for crude oil get high enough, shale may become profitable, and since it uses natural gas, even greater price pressure will mount for natural gas. Some refineries will either be redesigned or new ones will be built to use heavy oil as an input. The dire situation will probably mean environmental rules would be suspended in order to ease the transition to these new, unpleasant, raw resources.

We would also begin to see changes in the type of automobiles purchased. Smaller and more efficient cars would be dominant, with hybrid and pure electric vehicles well represented on roads. Public transportation would be used more, though this would probably be limited in the US to the East Coast and parts of the Midwest, where cities are concentrated enough to warrant its use. Industries would seek to minimize their energy use, and some of the heavy users would probably not survive. For the USA, these would include most of the steel industry (which is weak to begin with) and much of the automotive and appliance industries. Other industries, where there is high value relative to energy costs, would do well. For the USA, these include the electronics and computer software industries. The USA and other oil importing nations would have their economies transformed from heavy industrial activity to

5.1. A FIRST LOOK - THE FUTURE AS YOU SAW IT

service and intellectual property trades. It follows that oil companies which have substantial reserves on their balance sheets would be extraordinarily profitable, as would most energy related businesses with raw energy assets in their possession.

International trade would become more difficult, as transportation costs would rise, creating a natural barrier to trade. There are still many goods that could make the trade journey - those with high value relative to transportation costs. Heavier, low value goods would probably no longer trade over large distances. Food and grains would fall under this classification, and some food-short nations may well end up going hungry. Textiles would survive, but many nations would begin to search for closer countries to trade some goods with. This would lead to a greater concentration of geographically localized trade; it would push nations toward more trading in blocs. The shipping industry may do well, though with less total shipping, they would need to benefit from higher margins.

Simultaneously, remarkable financial transformation will occur. As crude oil prices rise, oil-rich countries will enjoy a windfall of cash that has never been seen before. This would constitute the largest peacetime transfer of assets from one set of countries to another in world history. Oil-rich countries would go on a spending and investment spree, buying both goods and assets from other countries. The cross-border purchases would keep the world financial system in a sort of tenuous equilibrium for some time, though the inflationary effect of rising oil prices would begin to erode the structure of the world financial system. Given the current claim that current known proven oil reserves will run out in 2030-40, this should reach a crisis level by 2020, and the higher oil prices combined with the US trade and government deficits should spark a serious inflation problem. At the same time the European Community countries would experience this as well, excluding Great Britain and Norway, both of which are petroleum exporters. A strong worldwide recession would probably occur, and with the weak world economy a temporary fall in crude oil demand would moderate prices.

At or about the same time that all of this is happening, several forces would combine to temper crude oil prices further. First, thanks to the sustained high oil prices, the search for new oil reserves would intensify, and it seems likely that new reserves would be found. Second, light sweet crude alternatives would get quite a boost, with heavy oil refineries being built. Old, long since capped oil wells would come back online, with heavy oil being pumped out to supply the new refineries. Most major energy alternatives (wind, solar, and nuclear) would become competitive. This would be driven, in part, by rising energy prices. In addition, technological innovations may reduce the cost of these alternatives making them even more competitive. By 2030, energy prices will have stabilized at a new, higher level. Much less crude oil would be in use, but overall the energy demand would probably not be much different from the time before the crisis.

The problem is that the United States would enter 2030 in a much weaker financial condition than it was in 2000. The damage caused by a severe inflation earlier in the century would cause the country to have difficulty attracting foreign investment, especially for government bonds. Fiscal deficits would be especially difficult, as bonds financing them would require high interest rates, and the money-financing alternative would be inflationary. Why is this important? Because Social Security obligations would be reaching higher levels, but Social Security tax revenues would be declining due to a smaller workforce population. The program would face the real possibility of insolvency within the decade, so a major revision to it would be necessary. This would probably include sharply lower benefits with a much higher retirement age.

With higher transportation costs, the demographic effects on trade would be diminished. A mid-century opportunity for many African countries to become major labor goods exporters would be weakened or not appear at all. Of course some oil-rich African countries would be doing well, and South Africa with its gold mines would also be doing well as world inflation would drive up gold prices. China's rising middle class, now aging, would find a world with higher prices for many imported goods, and the desire to own a car, for example, would be diminished by the high cost of ownership. A similar problem would exist for India. With weaker mar-

5.1. A FIRST LOOK - THE FUTURE AS YOU SAW IT

kets, and a depressed labor market in particular, China would enter a period with a demographic challenge (recall they are missing 50 million girls) with a very stressed population. That does not bode well for either China or the countries in its vicinity.

By 2050, most of the effects of higher oil prices would subside, and the world economy would operate with a higher energy cost base, but also with new energy technologies that would both reduce the amount of fossil fuels used and the amount of total pollutants associated with their use. In many respects, this might be a more desirable world to live in; a world in which most energy sources emit less carbon and are from renewable sources. The agricultural grain industry would probably be in good condition, having a shakeout during the oil price increase in the prior decade, though the industry would have replaced petroleum distillates with something else during the period. Presumably these replacement chemicals would be less hazardous to the environment than the petroleum ones.

Somewhere around 2050-60, Africa's labor advantage should become great enough that it can overcome the higher transportation costs that the world now faces. The problem is that the advantage has arrived a bit late, and with the world's attention on crude oil prices too little attention has been paid to the demographic transition that has been occurring in Africa. Repeated civil wars across the continent would have decimated populations and driven the continent deeper into poverty. Their starting point for development would be much lower than it would have been earlier in the century, and it is hard to know if the opportunity would be taken advantage of. If the political situation could be stabilized, many countries could start off on the road to a long, slow development, probably starting in textiles. Given the initially weak position of the continent, a concerted effort by the rest of the world would be necessary to capitalize on the opportunity Africa would now have.

With US economic and political power diminished by the period of rising oil prices, there would be no clear single country to lead, and one might guess that the United Nations might actually function as the world's legislative body. The relative clout of the Persian Gulf nations would be larger, given their extraordinary wealth and ownership of assets worldwide. It isn't hard to imagine that an increasingly isolated Israel would become engaged in a war with countries nearby; this would surely result in unpleasant consequences for all involved. Without a single leading country, many historic rivalries could break out in open warfare throughout the rest of the century. With luck, none would be on a world scale.

In some ways, this world might be better, given the concern about carbon emissions. The high crude oil prices would force both conservation and the switch to alternative energies, both of which would reduce carbon emissions. Labor in well-developed countries would also fare better due to natural trade barriers that would arise from higher transportation costs. The switch in ownership of financial assets to oil rich countries would result in a more intensely competitive capital market worldwide, though it is not clear that this would result in a more even distribution of capital throughout the world. Well-developed countries would find their balance of payments under greater downward pressure from the capital account, and it would be more difficult for them to run trade deficits. This would also make it more difficult for governments to run fiscal deficits. It seems likely that for the remainder of the century most countries would find it desirable to reduce their levels of international trade (using tariffs and quotas) in order to reduce their exposure to other countries' economies.

The unfolding of the century in front of us, given a rising oil price regime, is not very pleasant, but certainly livable. No doubt you've noticed the lack of good-looking Australians running around the desert shooting people for gasoline. It is not quite that future. But things would certainly change. For some this would be a better world; this is especially true for those who own petroleum-based assets at the beginning of the century. If you believe that oil prices are going to keep rising, investment in any major petroleum resource makes sense. The suggested damage to financial systems worldwide implies that there are few other places in which to invest. Gold and other precious metals tend to fare well during such periods, but they are very risky and have no particular rate of return. Money markets allow you flexibility and lower risk, but the mid-century financial mess might make such an investment undesirable. Investment in foreign countries which are also dependent on foreign oil imports are not likely to do any better than domestic investments. Your best choice would be investment in the energy sector, perhaps in alternative energies, and making sure your home is as energy-efficient as possible.

Are you ready to consider an alternative future? The analysis presented in the last few chapters suggested a very different future. It is punctuated by short-term spikes in the price of crude oil, but the price, on average, stays fairly constant. Because energy prices are not the key driving variable, a very different dynamic takes over. In such a future, trade and demography take center stage. While energy prices remain important, if oil prices do not rise consistently, the international economics and demography are more critical to understanding what your future might look like.

5.2 A new future: trade in an aging world

The most immediate issue for the United States, and for the world, relates to trade. This is operating at two levels: the demographic level and the energy (and other resources) level. The easy place to start is with the resource story.

Virtually every country in the world trades with others. There are many reasons for this, but the most obvious is the variation in raw materials in the world. Oil is a nice example, since it is clearly available in some countries and not (or not enough of it) in others. The Persian Gulf countries have a large amount of known proven reserves of oil, while the United States has a large amount needed. Saudi Arabia trades (exports) oil to the United States, where the USA is substituting its lack of resources with trade.

It is easy to see how trade substitutes for lack of oil, but it is much harder to see this process for resources like labor. Some countries like the USA have a large amount of capital, and its labor tends to be highly skilled and well educated. This makes US wages high, but means jobs that pay those wages need to be very productive. The US lacks cheap labor resources - those labor resources that use little skill or education and tend to be less productive. There are many labor-rich countries in the world; these are countries with a large low-skilled labor force. China comes to mind, and sure enough, China tends to export those things that utilize low-skilled labor resources. The USA imports these goods to substitute its lack of low-skilled labor.

That's not a very dynamic story, so our attention needs to turn toward how this changes over time. The key driving force here is demographic change. In those societies that are currently major exporters of low-skilled labor goods a change will be taking shape in the first third of the century. China, India and others are aging. Their population structure will be getting older. Because their age distribution will be flattening, the concentration of those who are of working age will rise in their societies. The good side to this is that their average productivity will rise, in part due to an increase in the average experience level of their workers. This also means that the average wage level in these countries will begin to rise. The bad side of this is that the price of their low-skilled goods will begin to rise. This leaves market space for other countries to compete with them using their low-skilled labor force. If these low-skilled labor countries are unable to develop these "light industries" then the price of these goods will rise.

Who has these low-skilled labor forces in 30 years? There are several countries around the globe that will have such a labor force - Bolivia, possibly Peru, a few island nations in Oceania, perhaps a few central Asian countries like Bangladesh and Nepal - but the highest concentration of these countries will be in Africa.

5.2.1 Will the 21st century be Africa's?

Africa is the great question of this century. After a half-century of civil wars, hunger and disease, the nations of this continent may be able to have enough political stability to build their economies. South Africa has already achieved this and will probably remain the wealthiest economy in Africa. But central Africa and much of northern Africa are still in a very weak position to take advantage of any economic opportunity that may come their way. There is so much at stake for both the Africans and the rest of the world that something needs to be done to assist them.

5.2. A NEW FUTURE: TRADE IN AN AGING WORLD

The continent as a whole needs political stability. There is no obvious means to achieve this. As a general rule, economies do not do well during civil wars, and much of the continent has had either open civil wars or violent civil unrest for much of the past half century. Financial assistance and access to well-developed countries' markets would probably help. If the world can give Africa an opportunity to grow in whatever way each country chooses, there is a chance for stability.

The best reason for the world to do this occurs by about 2030. When the demographic shifts occur in China, and much of the rest of the currently rapidly developing countries of the world, Africa will be the last place with a large low-skilled labor force. There are a host of manufactured goods that need this sort of labor supply, most prominently textiles. A smooth transition of these manufactures to labor-rich countries would provide a path for the world economy to grow in a balanced way. It would also provide Africa a stable means of economic development, one that they could build on for their future.

Recall that it has been suggested that international trade may have reached its limit, and we should expect that openness to international trade and levels of trade in general will decline in the world. It has also been suggested that trade may break into blocs, and Africa does not enjoy access to any major bloc in such a structure. Since Africa's opportunity will come in the 4th and 5th decades, it sounds like the continent is doomed by poor luck. That might not be the case.

The strongest reason for countries to trade with each other is some inherent advantage each has with reference to a particular set of goods and services. Africa's nations will have a compelling advantage in their labor supply.

Should this future come to pass, things get more interesting by 2070. Incomes in Africa should begin to rise, and the typical decrease in the birth rate and increase in average age of the population should follow. How China deals with this in the next 30 years should shed some light on what we can expect to happen in much of Africa. With the right mix of economic opportunity and political openness, African countries can make the transition without wide-scale civil wars. Whether that is possible remains to be seen. Nonetheless, by 2100, there may not be any place in the world with a large supply of low-skilled labor, or for that matter a high concentration of young people relative to other ages. Africa, and the world as a whole, will be older, and with any luck wiser.

5.3 The environment, energy and technology

The likely sharp, though short-lived spikes in energy prices at various times in the future may well spur energy conservation measures. This is a demographic effect. Since we know that most Western societies will have a large, older, and retired population, short-term sharp spikes in prices have a strong effect on their real (inflation adjusted) income. They are more sensitive to price changes because they are on a fixed income. So even if it turns out that average energy prices remain constant for much of the century, variation in those prices still have an effect. Most of the effect comes from the demand side, through either substitution or conservation measures to evade the higher prices.

How might this look? Consider a recent innovation - LED lights. These have been around for quite awhile, in use for computer lights and other indicator illumination. Recently though, they have been applied to use in flashlights. They last considerably longer than incandescent bulbs, and because they use less electricity to give the same output, the batteries last longer. They are now being applied to Christmas lights. These light strings also have longer bulb life and use less electricity, about 80% less, than incandescent bulbs. They are more expensive than standard light strings, but if there were an electricity price spike, even if temporary, people will tend to buy these energy saving lights. When the price of electricity falls back to normal levels, people still have (and will use) the energy saving bulbs. Short-term price increases in energy can cause long-term declines in demand, through conservation.

There is another good example regarding this idea: gasoline

5.3. THE ENVIRONMENT, ENERGY AND TECHNOLOGY

conservation. Again, even if the average price of gasoline declines (in real terms) over the century, there is evidence that its price volatility will be high. During price spikes people switch to smaller cars to conserve fuel and fuel costs. When the price returns to normal, some will buy a larger car, but those on fixed (and low) incomes cannot afford to do this - they keep the small car, and continue to conserve fuel. A short-term price increase results in demand decrease, through conservation.

This sort of response to price movements should be expected in the coming century. Still, the question remains, can innovations such as LED bulbs save us from possible energy shortages or environmental damage? The question has no clear answer, but it can be reformulated to ask: What innovations will be necessary to allow the economy to keep moving forward even if the resources of the planet begin to run short? We can't know the specific technologies that do this, but we can surmise the general areas where these innovations are likely needed.

Energy is such a broad category, with so many applications, that it is clear that innovation can be a significant part of the future. It is clear that increases in energy efficiency are practical, and will remain a major category for the foreseeable future. More efficient engines, heating and cooling devices as well as lighting are all areas in which we should expect improvement. In the same vein, we should observe advances in insulation and temperature control to decrease energy consumption. These areas are important, but tend to be mundane for both investors and news.

The larger story will probably revolve around fuel sources. There are at least two reasons that this is important: the possibility that readily accessible and useable energy sources are declining (like crude oil and natural gas), and the sense that the environmental effects of our use of fossil fuels are damaging. While it is arguable how important these two factors are, it is an argument about degree of importance, not that these things are facts. That means that these issues will be with us for quite some time. It also means that there is market space for profitability for anything that may alleviate these problems. We can be nearly certain that fuel sources will be a significant part of how the 21st century develops.

It was already mentioned in the section on oil history, but the idea of significant price variation in oil and other fossil fuels is important. Whenever a resource price is high, there are incentives to find more. The search for crude oil is no exception. Crude oil was originally found in seeps, and in oily sandy deposits that required no drilling; oil wells were effectively dug by hand. That made oil exceptionally easy to find. We are probably past that stage, though modern geology technology makes finding oil without the benefit of obvious seeps at least possible. So why are we not finding more oil? There's a corollary to Murphy's Law: whatever you are looking for, it will be in the last place you look. The planet is a very large place, and we not only have to look for oil, but there needs to be an incentive to look for oil. It is clear from past oil price data that there are not always strong incentives to go looking. Any period of sustained high crude prices will mean an intensified effort to find more. That does not mean that we will find more, but we certainly will not if we are not looking for it. The bad news here is that to make looking for oil desirable, there has to be a period of relative scarcity - shortages if you will, that drive oil prices up. These high prices will push exploration, and more than likely will result in newly discovered reserves. Given the need for price movements, and possible environmental damage from fossil fuel use, the world may prefer an alternative fuel source.

What is needed to create a big advance in fuel sources? Think about what makes gasoline so useful. It has great energy compactness (this is often called energy density) - in a relatively small space it contains a lot of energy. It has great portability. It is easy to carry and deliver through pipelines and trucks. It is easily measured with a fair degree of accuracy. And we have a large number of refineries in place that can convert crude oil to gasoline. How can this be replaced (or at least substituted in part)? The most critical aspects are the large energy in a small space factor and gasoline's portability. This could be substituted with a very good means of storing electricity - a very good battery. Provided you could store enough electricity with batteries, you could drive a car for 400 miles or more without a charge. You could also run a machine or electronic device in remote places. Electricity is

5.3. THE ENVIRONMENT, ENERGY AND TECHNOLOGY

already distributed throughout the world, and with a little effort charging stations could make longer drives realistic. Rapid charging would also probably have to be possible, as it is doubtful that people would be willing to wait overnight to finish their trips. The technological hurdles in this are high, but the profit incentive is also high.

It has already been mentioned that the use of heavy oil also seems to be practical. Should light sweet crude supplies run short, there can be little doubt that heavier crude oils will come into use. This will require new refineries that can handle this type of oil. While it is known that Venezuela has a large proven reserve of this type of oil, it is likely that many old wells in the United States as well as other countries also have such oil available. These wells were capped years ago because there was no application for heavier oils when light sweet crude remained cheap. It is not known how much of this type of petroleum is available in the world. The use of heavier oils also comes with a cost: it is likely that there will be unpleasant environmental effects resulting from its refinement.

What about fuel cells? The process involves burning hydrogen with oxygen, both providing a great deal of energy and a pleasant by-product - water. The process can use electricity to power the original separation of the two elements, so it should be fairly easy to produce. Can't this work? There are two problems here. First, it takes a fairly large amount of energy to make the fuel cell. The energy that comes back from the fuel cell is much less, so the process is inefficient, at least at the present time. But portability is paramount. Gasoline is volatile; it is explosive, which is one of things that makes it so useful. Hydrogen is much more dangerous stuff. It is also a gas at room temperatures, so it has to be cooled and compressed. While it is possible to contain it, the thought of societies with millions of automobiles and millions of machines all over the planet, needing to be carefully maintained else risk severe explosions, does not seem to be a place in which any of us want to live. In mass use, fuel cells are not likely to be practical.

Simply replacing gasoline with electricity won't solve the energy problem facing us in the 21st century. First, some applications for fossil fuel can't be readily replaced. The most obvious is jet fuel. Jets burn a very high-octane (a very refined) version of gasoline. There are no clear replacements for jet fuel that could be used to the degree it is used now. Rocket fuel, which is not based on fossil fuels, is probably not practical in jet applications. You simply cannot provide jet-level thrust with an electrical motor. There is no foreseeable alternative to jet fuel, though it is possible something will be developed. Given the eminent practicality of jet travel, and the lack of alternatives to jet fuel, it seems likely that jet fuel prices will rise to the point that it will be provided to airlines and other users. That, in turn, will push up prices for other oil-refined products, providing an incentive to replace gasoline, heating oil, and oil's use in insecticides and other products with some other alternative. Should this happen, there will be a period in which jet travel will be somewhat more expensive than it is now.

The use of refined crude oil products in insecticides, pesticides and fertilizers will need to be replaced with something else. Given the widespread use of these is in agriculture, it seems likely that some renewable and agriculturally-related product will become the replacement. With any luck, the product will also be more environmentally friendly. Again, this will be driven by rising crude costs, possibly temporary, but ultimately providing market space for new alternatives to become profitable.

The fact is that energy demand will rise over the century. A rising world population makes that certain. The speed by which this demand rises depends largely on how rapidly incomes in the world rise. As of 2004, the average per capita world output growth rate is about 2%. All things considered, that's a fairly fast rate, effectively doubling the world's output in around 40 years. Should that rate of growth continue to hold, energy demand will approximately double as well. How that demand is satisfied is critical. The Kyoto Protocol, an agreement between most countries in the world (but not the United States) has specified limits on industrial and other carbon emissions. The effect of the agreement should push countries toward employing alternative energy sources like wind and solar, even if crude oil prices do not rise. There are serious doubts that these alternatives can satisfy the increase in energy demand over the century, let alone decrease the world's demand for fossil fuel energy. However, if countries stay with the Kyoto Protocol, there should be market incentives to develop new energy alternatives as well as expand the use of current technologies.

5.4 A fall in the US dollar in our future

The US dollar acts as a store of value for most countries in the world in 2005. Their central banks use it as their primary foreign reserve holdings and their trade currency. There are good reasons to believe this will not last throughout the century. Before you sell all of your US dollar holdings and buy all things in Thai bhat or Malaysian ringit, there are a number of things you should know.

The strength of the US dollar is part of the reason that the United States has run such large trade and fiscal deficits for the last two decades of the 20th century, and the first decade of the 21st. It is not likely that the world will continue to finance US deficits for much longer. How much longer? That's hard to say, but there should be some warning signs that the world is about to limit the US credit supply.

The United States currently enjoys the top leadership role for the world's political and military institutions. It also has the world's largest single economy. These things together are enough to mean that the USA is the investment destination of choice whenever the world seems risky. So what we should look for is something that changes this current status. What might do that?

The European Union (EU), once fully formed, should rival the US economy in size, perhaps even exceed it. If the European Economic and Monetary Union (EMU), (i.e. the countries that use the euro as their domestic currency) should expand to cover all EU countries, then the euro in effect represents an economy as large or larger than the US dollar. Given that England has not joined the EMU, and that other countries may not be allowed to join, it is possible that this may never happen. Just the same, the euro does represent a large joined set of economies, and may be competitive with the dollar as a reserve currency. If the US ever allows for sizable inflation, and the euro remains stable, a shift of reserve currency by the world's central banks would be virtually certain.

Large and continuous trade and fiscal deficits are only part of the problem here. Every time crude oil enjoys a price spike, the enormous demand for foreign oil in the USA means an extraordinary amount of US dollars flood foreign markets. These spikes are likely to occur during any Middle East instability or during unusual demand for oil by rapidly developing countries. For example, China's sharp increase in the demand for oil from 2000-2005 appears to be a result of weakness in the electricity supply infrastructure, which led factories to use diesel generators to insure their own electricity supply. The Iraq war also fed the higher prices. The Iran crisis of 1979-80 drove oil prices to their highest real level yet. Every one of these events meant that US dollars flooded foreign markets, and how those dollars are used very much affects whether the US gains, loses or stays even with foreign investment. Not only are new conflicts in the Middle East possible, but even problems with royal succession could cause instability, and thus spikes in oil prices. There are currently (2005) about 20 brothers who are in line to become to king of Saudi Arabia. Their sons, who are the next generation of kings, number in the hundreds. That is surely a recipe for instability.

In the first few decades of the 21st century we should see repeated ups and downs of the US dollar, though not necessarily based on a regular cycle. Very large US fiscal and trade deficits should make such swings quite wide. If the US were to get control of its deficits, these swings should be less volatile. By the third and fourth decades of the 21st century, the demographic issue kicks in. Social Security may well be in or near insolvency, and the amount of new incoming funds will be unable to solve the problem, because there will be fewer working-age people relative to those who are older and presumably retired. Because those who are older tend to vote, they will exert political pressure to finance Social Security through the ordinary fiscal budget. Either taxes will go up, or a new fiscal deficit will emerge, and one that cannot be removed easily. If the deficit can't be financed with bonds (some of which are purchased by foreigners), then it will be financed by expanding the money supply. That is a recipe for inflation.

By 2050, with the United States dealing with a serious and ris-

5.4. A FALL IN THE US DOLLAR IN OUR FUTURE

ing inflation rate, things really get unpleasant. First, the dollar price of oil would rise, as would all imported goods, since the dollar is losing its value over time. Most bonds, those that are not inflation-indexed, would rapidly lose their value over time. Stocks also do not tend to do well during inflations. One of the few good places to be is in some foreign markets where the currency is not inflating. That's true for domestic (US) residents, and more importantly, for foreigners. That means a withdrawal of funds from the US and investment elsewhere. The tide of currency leaving the USA would be more like a tsunami, and the US dollar would fall very sharply. Most central banks in the world would drop the US dollar as a reserve currency, which would push the US dollar further down. With the US dollar in a tailspin, the world economy would likely contract quickly. That is a recipe for disaster.

All of this can happen without the USA losing its preeminence as the world political and military leader. It can also happen without the price of oil steadily rising as some have predicted, though the occasional price spikes in oil suggested here create part of the inflation problem. Either a decline in the prestige of the United States or a prolonged increase in the price of oil could cause a major financial disaster like the one just suggested. However, it is hard to foresee a major adversary arising to challenge the current status of the USA in the next 50 years. The data on crude oil also does not suggest a steady increase in oil prices. And yet, without either of these troubling scenarios, the USA and the world can still have a financial crisis of large magnitude. Surely something can be done to avert this.

This mid-century horror need not await us. There are things we can control and things we can't control. The demographic facts are already known. We can't change them. But large deficits, at least large fiscal (government) deficits can be changed. It is probably best to aim for a near balanced budget over the business cycle, which in the USA is about 5-7 years in length. If the United States plans properly for Social Security now, it can probably avoid its insolvency in 30-40 years. That means somewhat higher taxes now, and some plan to increase the retirement age. By 2040, the US will need people over 65 to work anyway, it might as well work on how this makes sense now. If the United States can enter the flattened age structure that it will have in 30-40 years with sound fiscal policy, light deficits and a functional retirement system, it should be able to weather most financial and economic storms. That is a recipe for success for the rest of the century.

5.5 Where to find other views of the future

Perhaps you are not satisfied with the view of the 21st century presented here. There many alternatives, often centered on different issues with very different things to say. Here are some suggestions for your consideration.

First, let me suggest you watch a movie: Mad Max 2: The Road Warrior (1981), a Metro-Goldwyn-Mayer motion picture, starring Mel Gibson. It's the film I have referred to as "having good-looking Australians running around shooting people for gasoline." It's a fairly good film, and lots of fun. It most certainly provides a very different view of the future than I have suggested.

Thomas L. Friedman concentrates on communication technology in *The World Is Flat*. Published in 2005 by Farrar, Straus and Giroux, the book explores how the Internet and other telecommunication technologies are changing the competitive landscape for people all over the world. His discussion will provide you a better understanding of outsourcing and exchange of services on a global basis. He's not alone in the idea of technology driving world change, and a more complete picture might be found in *The Information Age: An Anthology on Its Impact and Consequences* Edited by David S. Alberts and Daniel S. Papp, published by the National Defense University in 2002. It is a collection of papers by a wide variety of authors that explore technology's impact on government, the military, business and international relations. I'm not so sure that these changes are the driving force throughout the century, but you'll have to judge that for yourself.

You could also take a look at the World Future Society homepage, at http://www.wfs.org/ which includes links to The Futurist magazine, and reviews of books about the future. Resources for the Future, at http://www.rff.org/ is a policy center with numer-

5.5. WHERE TO FIND OTHER VIEWS OF THE FUTURE

ous research areas. It would be a good place to go to see specific topics about the future. The Millenium Project, which is based at the American Council for the United Nations University is another site with an array of topics. You can find at http://www.acunu.org/ on the internet.

The website for this book is at http://www.centuryforecast.com. The links suggested in the book as well as enhanced information and new links will be available. The site will also offer greater technical depth in the econometric analysis that supports the material in the book. The site will be updated periodically to keep up with changes in the future.